Lecture Notes in Control and Informatio[n Sciences] 1

For further volumes:
http://www.springer.com/series/642

S.R. Shimjith, A.P. Tiwari, and B. Bandyopadhyay

Modeling and Control of a Large Nuclear Reactor

A Three-Time-Scale Approach

Authors
S.R. Shimjith
Research Scholar
Indian Institute of Technology Bombay
Scientific Officer (F)
Bhabha Atomic Research Centre
Mumbai
India

B. Bandyopadhyay
Professor
IDP in Systems and Control Engg.
Indian Institute of Technology Bombay
Mumbai
India

A.P. Tiwari
Scientific Officer(H+) and Professor
Bhabha Atomic Research Centre
Mumbai
India

ISSN 0170-8643
e-ISSN 1610-7411
ISBN 978-3-642-30588-7
e-ISBN 978-3-642-30589-4
DOI 10.1007/978-3-642-30589-4
Springer Heidelberg New York Dordrecht London

Library of Congress Control Number: 2012940248

Printed on acid-free paper

Springer is part of Springer Science+Business Media (www.springer.com)

OM SAHANA VAVATU SAHANAU BHUNAKTU
SAHA VIRYAM KARAWAVAHAI
TEJASVINAVADITAMASTU
MA VIDVISHAVAHAI
OM SHANTI SHANTI SHANTI OM

(Together may we be protected,
Together may we be nourished,
Together may we work with great energy,
May our journey together be brilliant and effective,
May there be no bad feelings between us,
Peace, peace, peace.)

Kato Upanishad

To our Parents and Teachers, who made us what we are now

Preface

Control and operation of large nuclear reactors pose several challenges. First and foremost difficulty is the development of a suitable mathematical model, which in itself is the most challenging task. Nuclear reactors of small core size are adequately represented by the well–known point kinetics model which depicts the dynamics of the core averaged neutron flux / power and the associated delayed neutron precursor groups. In large nuclear reactors, the flux shape may undergo variations which the point kinetics model is unable to capture. Hence for large nuclear reactors, it is essential to adopt more detailed space–time kinetics models for analysis. However, such models often will be of very large order which hamper their direct use in controller design applications. Hence it becomes essential to develop a suitable model which captures all the essential properties of the reactor core, at the same time having a trade–off between the model accuracy and model order. Furthermore, in most of the control related applications, it would be preferable to cast the model equations as a set of first order ordinary differential equations so that they can be expressed in standard state space form. This monograph introduces a promising candidate for space–time–kinetics modeling of large nuclear reactors which yields a simple but reasonably accurate mathematical model amenable for control related studies.

It is well known that physical systems like the nuclear reactors exhibit simultaneous dynamics of different speeds. For example, subsequent to an almost instantaneous change in the fission power, some variables like coolant temperature and concentration of certain delayed neutron precursors change in few seconds, whereas concentration of certain other fission products like xenon *etc*. take few hours to reach their respective new equilibrium values. Such a behavior essentially leads to mathematical models exhibiting multiple time scales, which are susceptible to numerical ill–conditioning. Hence it becomes essential to investigate multi–time scale properties of the model developed, and adopt suitable model order reduction or model decomposition techniques to address the ill–conditioning effects. Hence, this monograph further introduces novel techniques for decomposition of multi–time–scale systems and their composite control based on both state as well as output feedback, with emphasis on three–time–scale systems.

The authors would like to express their deep sense of gratitude to their parents and teachers who have made them capable enough to write this book. The authors wish to thank many individuals who had helped them directly or indirectly in completing this monograph. In particular authors wish to express their thanks to their colleagues in Reactor Physics Design Division and Reactor Engineering Division, BARC. Finally authors wish to acknowledge the support, patience and love of their wives and children during the preparation of this monograph.

S.R. Shimjith
BARC, Mumbai, India *A.P. Tiwari*
IIT Bombay, Mumbai, India *B. Bandyopadhyay*

March 2012.

Contents

Acronyms

AHWR Advanced Heavy Water Reactor
AR Absorber Rod
FOS Fast Output Sampling
PHWR Pressurized Heavy Water Reactor
RR Regulating Rod
SDS Shutdown System
SR Shim Rod

Glossary

A System matrix.

B Input matrix.

$\tilde{C}_{i_h}$ Concentration of ith group of delayed neutron precursor.

C_{i_h} Normalized concentration of ith group of delayed neutron precursor.

D_1 Fast group diffusion coefficient.

D_2 Thermal group diffusion coefficient.

E_{eff} Average thermal energy liberated per fission.

E_Z Identity matrix of dimension Z.

H_j Position of the jth regulating rod.

I_h Iodine concentration in node h.

J Quadratic performance index.

K State feedback gain matrix.

L Fast output sampling gain matrix.

M Output matrix.

P Pressure.

$\hat{Q}_h$ Reference power in node h.

Q_h Fission power in node h.

Q_L Power generated in left half of the core.

Q_R Power generated in right half of the core.

Q_{q_i} Power generated in i^{th} quadrant of the core.

Q_{1_h} Fraction of power in node h available for bringing the coolant to boiling point.

Q_{2_h} Fraction of power in node h available for boiling.

R_h Ratio of fast flux to thermal flux in node h.

T Transformation matrix.

V_{b_h} Volume in node h where boiling occurs.

V_h Volume of node h.

V_{C_h} Volume available for coolant flow in node h.

V_s Volume of steam in steam drum.

V_w Volume of water in steam drum.

X_h Xenon concentration in node h.

Z Number of nodes.

h_c Latent heat of vaporization.

h_d Enthalpy of downcomer water.

h_s Enthalpy of saturated steam.

h_w Enthalpy of saturated water.

h_s Enthalpy of saturated steam.

h_w Enthalpy of saturated water.

k_{eff} Effective multiplication factor.

m_d Number of delayed neutron precursor groups.

q_{d_h} Coolant inflow rate to node h.

q_d Net downcomer flow rate.

q_f Feed water flow rate.

q_{r_h} Coolant outflow rate (riser flow) from node h.

q_r Net riser flow rate.

q_s Steam flow rate.

v_1 Speed of fast neutrons.

v_2 Speed of thermal neutrons.

x Average core exit quality.

x_h Coolant exit quality of node h.

u Input vector.

y Output vector.

z State vector.

z_{f1} State vector of 'fast 1' subsystem.

z_{f2} State vector of 'fast 2' subsystem.

z_s State vector of 'slow' subsystem.

α_{hk} Coupling coefficient characterizing neutron leakage from node h to node k.

$\overline{\alpha}_h$ Average coolant void fraction along the axis in node h.

β Average fractional delayed neutron yield from thermal fissions.

β_i Average fractional delayed neutron yield of ith group of delayed neutrons.

δ Deviation from equilibrium value.

$\triangle_{hk}$ Centre–centre distance between nodes h and k.

ε, μ Small positive parameters.

γ_I Average fractional yield of ^{135}I from thermal fissions.

γ_x Average fractional yield of ^{135}Xe from thermal fissions.

λ_i Decay constant of ith group of delayed neutron precursor.

λ_I Decay constant of ^{135}I.

λ_x Decay constant of ^{135}Xe.

ℓ_h Average prompt neutron life time in node h.

ϕ_1 Group 1 (fast) flux.

ϕ_{1_h} Group 1 (fast) flux in node h.

ϕ_2 Group 2 (thermal) flux.

ϕ_{2_h} Group 2 (thermal) flux in node h.

ϕ_h Effective group flux in node h.

ρ_h Reactivity in node h.

ρ_{h_α} Reactivity feedback due to void fraction in node h.

ρ_{h_f} Reactivity feedback due to thermal hydraulics.

ρ_{h_u} Reactivity contributed by the regulating rods in node h.

ρ_{h_x} Reactivity feedback due to xenon in node h.

σ_{ax} Microscopic absorption cross–section of ^{135}Xe.

$\overline{\sigma}_{ax}$ Normalized microscopic absorption cross–section of ^{135}Xe.

ρ_d Density of downcomer water.

ρ_s Density of saturated steam.

ρ_w Density of saturated water.

ω_{hk} Effective group neutron leakage coefficient.

ω_{1hk} Fast neutron leakage coefficient.

ω_{2hk} Thermal neutron leakage coefficient.

ν Average total neutron yield from thermal fissions.

φ Eigenvalue.

Σ_{a1} Macroscopic absorption cross–section of fast neutrons.

Σ_{a2} Macroscopic absorption cross–section of thermal neutrons.

Σ_{a_h} Homogeneous macroscopic absorption cross–section in node h.

Σ_{f1} Macroscopic fission cross–section of fast neutrons.

Σ_{f2} Macroscopic fission cross–section of thermal neutrons.

Σ_{f_h} Homogeneous macroscopic fission cross–section in node h.

Σ_{12} Macroscopic scattering cross–section of neutrons (fast to thermal).

Σ_{12} Macroscopic scattering cross–section of neutrons (thermal to fast).

List of Figures

Chapter 1
Introduction

Atoms, the basic component of all matter, are incredibly tiny. But stored within the nucleus at the centre of each atom is the most powerful form of energy known. In 1938, Otto Hahn and Fritz Strassmann discovered that bombarding the nucleus of the uranium atom with neutrons changed some of the uranium into barium. In 1939, Austrian physicists Lise Meitner and Otto Frisch recognized Hahns experiment as the splitting of the uranium atom, and named the process as nuclear fission. During the same year, Jean Frederic JoliotCurie, Hans von Halban, and Lew Kowarski in France found that, on average, each fission of ^{235}U releases about 2.5 additional neutrons. The discovery made apparent the possibility of a chain reaction. Subsequently in 1942, a group of noted physicists, headed by Enrico Fermi, directed the construction of the worlds first successful nuclear reactor, at the University of Chicago in Illinois. It produced the first artificial chain reaction. Subsequent to these historical events, scientists learned how to release the nuclear energy and put it to use. Among the peaceful uses of this form of energy, is the production of electricity. A facility designed to convert nuclear energy into electricity is called a nuclear power plant.

In the 30 countries that have nuclear power plants, the percentage of electricity coming from nuclear reactors ranges from 78 % in France to just 2 % in China. As of March 2008, there are 439 nuclear power plants around the world, while 35 more are under construction. The USA has the most with 104, France is next with 59, then Japan with 55 and Russia has 31 and seven more under construction. The expansion in nuclear power generation is centered in Asia. A total of 20 of the 35 plants under construction are in Asia, while 28 of the last 39 plants connected to the grid are also in Asia. The reactors in todays nuclear power plants are of various designs. However, the trend now is to build nuclear reactors of large capacity which can be operated with relatively uniform flux distribution. It minimizes the per unit electricity generation cost.

Because of the potential for accidents or sabotage at nuclear power plants, the operation and control of these plants represents a complex problem. Several safety and control features are engineered at the design stage and operational policies are incorporated to avoid accidental release of radioactivity to the general population. The problems are further complicated in case of large nuclear reactor.

S.R. Shimjith et al.: Modeling and Control of a Large Nuclear Reactor, LNCIS 431, pp. 1–15.
springerlink.com

The physical dimensions of large nuclear reactors are typically several times large compared to the neutron migration length so that any deliberate attempt to operate the reactor with flattened radial and axial flux distributions coupled with xenon poisoning effect leads to complex operational and control problems. A small fraction of ^{135}Xe is formed directly in fission but the major portion results from the radioactive decay of ^{135}I, with a half life of 6.7 hours. The thermal neutron absorption crosssection of ^{135}Xe is very large and it undergoes radioactive decay at a relatively slower rate than ^{135}I does. Hence, the immediate effect of an increase in neutron flux in the reactor is to cause a decrease in the concentration of ^{135}Xe, which in turn results into an increase in neutron flux. This continues for about 10 hours when the reverse starts taking place. Production of ^{135}Xe through the decay of ^{135}I becomes appreciable and the neutron flux starts decreasing due to increase in thermal neutron absorption by ^{135}Xe, resulting into further build up of concentration of ^{135}Xe. In this manner, oscillations of the neutron flux are introduced by ^{135}Xe. Such oscillations induced by ^{135}Xe can be broadly classified in two categories - fundamental mode or global power oscillation and higher mode or spatial oscillations. Generally speaking, global power oscillations can be readily noticed and suppressed by control system while spatial oscillations are not. Here the global reactor power can remain constant so that no change in the coolant outlet temperature is effected, the oscillations being in spatial distribution of power inside the core.

If the oscillations in the power distribution are left uncontrolled, the power density and the time rate of change of power at some locations in the reactor core may exceed the respective permissible limits, resulting into increased chances of fuel failure. Hence, in large thermal reactors it becomes necessary to employ automatic power distribution control systems besides the system for control of global power output. Such systems have been provided in large pressurized water reactors, advanced gas cooled reactors and pressurized heavy water reactors. In general, the control system consists of a number of appropriately located reactivity controllers which are manoeuvred on the basis of the flux-distribution signal derived from readings of several incore detectors. The objective is to maintain the core power distribution close to a desired shape.

The foremost problem encountered in control analysis and design of a large reactor is the model development. Nuclear reactors of small and medium size are generally described by the point-kinetic model which characterises every point in the reactor by an amplitude factor and a time independent spatial shape function. This model is, however, not valid in case of large reactors because the flux shape undergoes appreciable variation with time. Explicit consideration of the variation of the flux shape becomes necessary. Three-dimensional finite difference approximations of one or multigroup diffusion equations are generally employed in calculation of the power distribution inside the reactor core. However, their application to control problems is not amenable. Some approximation based on nodal, modal or flux-synthesis methods becomes necessary to obtain a sufficiently accurate model of reasonable order.

It is well known that physical systems like the nuclear reactor exhibit simultaneous dynamics of different speeds. For example, subsequent to an almost instantaneous change in the fission power, some variables like coolant temperature and concentration of certain delayed neutron precursors change in few seconds, whereas concentration of certain other fission products like xenon *etc*. take few hours to reach their respective new equilibrium values. Such a behavior essentially leads to mathematical models exhibiting multiple time scales, which are susceptible to numerical ill–conditioning. Hence it becomes essential to investigate multi–time scale properties of the model developed, and adopt suitable model order reduction or model decomposition techniques to address the ill–conditioning effects.

The static output feedback is one of the most investigated problems in the control theory and applications [112] and it has been applied to PHWR control problem [40]. Conventionally, small and medium size nuclear reactors are generally controlled based on feedback of total power (or core averaged power), whereas large reactors may require feedback of spatial power distribution alongwith the total power feedback for effective spatial control. However, a major limitation of the static output feedback is that it may not guarantee arbitrary pole placement. Spatial control using state feedback would have been a better option for arbitrary pole placement. However, practical implementation of such a controller would require a state observer which will increase the complexity. The use of output information through modern control techniques such as periodic output feedback or fast output sampling would be better. However, in such applications the intricacies associated with multi–time–scale properties of the model need to be handled carefully.

1.1 Modeling of Large Nuclear Reactors

The central problem in reactor analysis is the determination of the spatial flux and power distribution in the reactor core under steady state as well as transient operating conditions. There is, however, a considerable variation in the degree of accuracy and spatial details of the power distribution required in different facets of reactor analysis and design. The fundamental equation governing the neutron flux distribution with time in a nuclear reactor is the Boltzmann transport equation but its exact solutions are available only in a few special cases and most of these are not time dependent. Fortunately, the neutron transport may frequently be represented by the following set of multigroup diffusion equations, in which the energy variable is discretized and the slowing-down process is represented by intergroup transfer cross-sections.

$$\mathbf{V}^{-1}\frac{\partial \Phi}{\partial t} = \left[\nabla.\mathbf{D}\nabla - \mathbf{A} + (1-\beta)\chi\mathbf{F}^T\right]\Phi + \sum_{i=1}^{m_d}\lambda_i\chi_i\tilde{C}_i, \tag{1.1}$$

$$\frac{\partial \tilde{C}_i}{\partial t} = \beta_i\mathbf{F}^T\Phi - \lambda_i\tilde{C}_i, \tag{1.2}$$

where $\boldsymbol{\Phi}$ is a column vector, each component being the scalar flux in an energy group, and $\tilde{C}_i$ is a scalar representing the precursor density for the ith delayed neutron group. The neutron speeds are represented by the diagonal matrix

$$\mathbf{V} = \text{diag.}\ [v_1\ v_2\ \dots\ v_G]$$

having one entry for each group. Similarly, the diffusion constants appear in the diagonal matrix

$$\mathbf{D} = \text{diag.}\ [D_1\ D_2\ \dots\ D_G]\,.$$

The matrix $\mathbf{A}$ representing the absorption and scattering processes is given by

$$\mathbf{A} = \begin{bmatrix} \Sigma_{a1} & 0 & \dots & 0 \\ 0 & \Sigma_{a2} & \dots & 0 \\ \vdots & \vdots & \ddots & \vdots \\ 0 & 0 & \dots & \Sigma_{aG} \end{bmatrix} - \begin{bmatrix} \Sigma_{11} & \Sigma_{12} & \dots & \Sigma_{1G} \\ \Sigma_{21} & \Sigma_{22} & \dots & \Sigma_{2G} \\ \vdots & \vdots & \ddots & \vdots \\ \Sigma_{G1} & \Sigma_{G2} & \dots & \Sigma_{GG} \end{bmatrix}$$

where the first part representing absorption is diagonal and the second part representing intergroup transfer is a full matrix. The fission crosssections appear in the vector

$$\mathbf{F}^T = \left[\nu\Sigma_{f1}\ \nu\Sigma_{f2}\ \dots\ \nu\Sigma_{fG}\right].$$

The emission spectra of prompt and delayed neutrons are also column vectors

$$\chi = [\chi_1\ \chi_2\ \cdots\ \chi_G]^T\,,$$
$$\chi_i = [\chi_{i1}\ \chi_{i2}\ \cdots\ \chi_{iG}]^T\,.$$

The above set of eqn. (1.1) and (1.2) can be solved by using finite difference methods to discretize the spatial variable. To be more precise the reactor core is broken up into a spatial grid or mesh. Then the multigroup diffusion equation is integrated over a typical cell, and standard sum and difference formulae are used to represent the terms in the equation. These numerical methods are of value in special situations, *e.g.*, analysis of simple reactor systems with a high degree of symmetry, and verification of test cases for approximation methods. Eventually it may be feasible to calculate any spatially dependent transient by straightforward finite difference methods. However, despite tremendous improvement in computer speeds and storage capacity, development of approximation techniques is favoured in case of control analysis and design problems.

The most important approximation method of relevance here is the prompt jump approximation or the zero life time approximation. The method is well understood in case of the point reactor model in which the dependence of spatial neutron flux distribution on time is completely ignored. To understand the prompt jump approximation, consider the differential equations of the point reactor, given as

$$\frac{dQ}{dt} = \frac{\rho(t)-\beta}{\ell}Q+\sum_i \lambda_i C_i \tag{1.3}$$

$$\frac{dC_i}{dt} = \frac{\beta_i}{\ell}Q-\lambda_i C_i \tag{1.4}$$

where ℓ is generally small. When $\rho < \beta$, the right hand side of (1.3) contains a large negative term $\frac{\rho-\beta}{\ell}Q$. In such a case the derivative $\frac{dQ}{dt}$ is often a small difference between a large negative number $\frac{\rho-\beta}{\ell}Q$ and a large positive number $\sum_i \lambda_i C_i$. Formally, a solution of Q in (1.3) of the form

$$Q = Q_1+\ell Q_2+\ldots \tag{1.5}$$

can be attempted, which yields

$$\frac{dQ}{dt} = \frac{dQ_1}{dt}+\ell\frac{dQ_2}{dt}+\ldots$$

Substitution of these expressions in (1.3) yields

$$\frac{dQ_1}{dt}+\ell\frac{dQ_2}{dt}+\ldots = \frac{\rho-\beta}{\ell}Q_1+(\rho-\beta)Q_2+\sum_i \lambda_i C_i+\ldots \tag{1.6}$$

It can be inferred from (1.4) that C_i are of the order $\frac{1}{\ell}$ with respect to Q, in steady state and during a slow transient. Assuming that the time derivatives are small, and equating terms of order $\frac{1}{\ell}$, we have

$$0 = \frac{\rho(t)-\beta}{\ell}Q_1+\sum_i \lambda_i C_i,$$

whence

$$Q_1 = \ell\frac{\sum_i \lambda_i C_i}{\beta-\rho}. \tag{1.7}$$

This function Q_1, which satisfies (1.3) whenever $\frac{dQ_1}{dt}$ is sufficiently small to be ignored, is identified as the power level in the prompt jump approximation. It is worth to note that Q_1 satisfies a system of ordinary differential equations whose order has been reduced by one. This procedure of expansion in powers of a small parameter, yielding a first approximation that satisfies a differential equation of reduced order, is the well known method of singular perturbations [26, 120].

Equating the terms of order ℓ^0 in (1.6), we obtain

$$\frac{dQ_1}{dt} = (\rho-\beta)Q_2,$$

which yields

$$Q_2 = \frac{1}{\rho - \beta}\frac{dQ_1}{dt}.$$

Using (1.7) to evaluate the derivative, we obtain

$$Q_2 = -\frac{1}{(\beta-\rho)^2}\left(\sum_i \lambda_i \ell \frac{dC_i}{dt} + \frac{\sum_i \lambda_i \ell C_i}{\beta-\rho}\frac{d\rho}{dt}\right).$$

With this, (1.5) becomes

$$Q = \ell \frac{\sum_i \lambda_i C_i}{\beta-\rho} - \frac{1}{(\beta-\rho)^2}\left(\sum_i \lambda_i \ell \frac{dC_i}{dt} + \frac{\sum_i \lambda_i \ell C_i}{\beta-\rho}\frac{d\rho}{dt}\right) + O\left(\ell^2\right) \tag{1.8}$$

which is an asymptotic series in powers of ℓ that diverges if $\rho \to \beta$ and has for its first term the prompt jump approximation. This treatment may also be regarded as a special case of a general method for approximating the solutions of different equations containing one or more very small time constants. Since Q_1 as given by (1.7) satisfies a system of differential equations of reduced order, its interpretation as the approximate neutron density satisfying (1.3) with dQ/dt omitted implies that a fast decaying transient part of the solution is missing from the approximate neutron density. The lost initial condition is replaced in this approximation by a requirement of the continuity of $(\rho-\beta)Q$.

By (1.7) the approximate neutron density in the one–delay–group case is

$$Q_1 = \frac{\ell \lambda C}{\beta-\rho},$$

from which if we solve for C and substitute the result into the delayed neutron precursor equation

$$\ell \frac{dC}{dt} = \beta Q - \ell \lambda C,$$

we obtain

$$(\beta-\rho)\frac{dQ_1}{dt} = \left(\lambda\rho + \frac{d\rho}{dt}\right)Q_1.$$

This is the first–order differential equation for the neutron density in the prompt–jump approximation.

Substituting the value of Q_1 from (1.7) in the delayed neutron precursors' equation (1.4), we have

$$\frac{dC_i}{dt} = \frac{\beta}{\beta-\rho}\sum_{i=1}\lambda_i C_i - \lambda_i C_i$$

which reduces to

$$\frac{dC}{dt} \simeq \frac{\lambda \rho}{\beta - \rho} C$$

if only one group of delayed neutrons is considered.

In principle, the prompt–jump approximation is valid if the second term of (1.8) is small. For one–delay–group, this condition is written alternatively as,

$$\beta - \rho \gg \sqrt{\left(\ell |\lambda \rho + \frac{d\rho}{dt}|\right)}.$$

In case of usual control related transients this criterion is generally satisfied as reactivity perturbations are usually small and they occur at much slower rates. Hence, prompt jump approximation serves as a useful tool in reactor dynamics analysis.

Application of prompt jump approximation to space-time reactor problems is however lacking although some attempts can be cited [17]. However, the major limitation of the above approximation, or even the direct use of the well known 'point kinetics model' defined by (1.3) and (1.4) is that they do not provide any information about the spatial power distribution inside the reactor core, whereas the central problem in analysis of large nuclear reactors is the determination of the spatial flux and power distribution in the reactor core under steady state as well as transient operating conditions. There is, however, a considerable variation in the degree of accuracy and spatial details of the power distribution required in different facets of reactor analysis and design. Basically, the behavior of neutrons in a nuclear reactor is adequately described by the time–dependent Boltzmann transport equation [24, 38], whose numerical solutions for reactor kinetics problems of practical interest are prohibitively difficult. So, approximate methods using the time–dependent group diffusion equations are employed. These methods can broadly be classified as space–time factorization methods, modal methods and direct methods [111].

Space–time factorization methods involve a factorization of the space and time dependent flux into a product of two parts. One part, called amplitude function, depends only on the time variable whereas the second part, called shape function, includes all of the space and energy dependence and is only weakly dependent on time [22, 39, 78].

Modal methods, on the other hand, utilize an expansion of the flux in terms of precomputed time–independent spatial distributions through a set of time–dependent group expansion coefficients [11, 27, 70, 106]. Another class of spatial methods called synthesis methods, which are almost equivalent to modal methods, are also prevalent. These methods use expansion functions that are static solutions of the diffusion equation for some specified set of initial conditions. Synthesis methods can often yield acceptable accuracy with a smaller number of expansion functions. However, selection of expansion functions for synthesis methods requires considerable experience.

Direct space–time methods solve the time–dependent group diffusion equations by partitioning the problem space into a finite number of elemental volumes, thereby obtaining spatially discretized forms of the coupled diffusion and delayed neutron precursors' equations. Direct methods are further classified as finite difference methods, coarse–mesh methods and nodal methods[111]. In each of these methods, the problem space is discretized by superimposing a computational mesh and the material properties are treated as uniform within each mesh box. Another method is the Finite Element Method (FEM) where the group flux is approximated as the sum of multi–dimensional polynomials that are identically zero everywhere outside some elemental volume, or as higher order polynomials thereof [48]. The FEM has advantages that it is not limited to regular mesh and that it is more accurate than finite difference method for large mesh. It is also reported that the FEM offer a common framework for all other direct methods and provide a general way of analyzing the convergence and error properties of these methods [34].

Application of FEM and finite difference methods requires a relatively fine mesh to ensure accuracy, which makes them computationally intensive. On the contrary, coarse–mesh methods assume that the reactor may be adequately described by a model consisting of homogeneous regions that are relatively large. Necessary reduction in discretization error is achieved through the use of higher order approximations to the spatial variations of the unknowns within a mesh box. In other words, determination of the multi–dimensional flux distribution within a mesh box is an integral part of the solution process [61].

All of these methods, in their fundamental form, target the reconstruction of accurate space–time dependent group flux distributions over the reactor core. In many applications it is also targeted to reconstruct flux distributions upto the fuel bundle or even the fuel pin level. However, minute details of such great extent are not required in control system studies, where the prime objective would usually be to model the total power and a coarse spatial power distribution within the reactor core with reasonable accuracy. In most of the control related applications, it would be preferable to cast the model equations as a set of first order differential equations so that they can be expressed in standard state space form. Among the various methods, nodal methods, from which simple first order equations governing node averaged power can be cast rather easily, form a better candidate under such scenarios. Moreover, with transient simulations being an integral part of the control system design and analysis task, nodal models further establish their relevance in control related applications. Over the period of time, nodal methods have been used extensively for the analysis and simulation of light water reactors and control system design of pressurized heavy water reactors [13, 85, 86, 113, 114, 117].

Like coarse–mesh methods, nodal methods also consider the division of the reactor core into relatively large, non–overlapping nodes. However, direct results of the solution process are often the node averaged fluxes. These methods generally demand additional relationships between the face averaged currents and the node averaged fluxes, often denoted as coupling parameters. The coupling parameters can be obtained from accurate reference calculations to relate the node interface averaged currents to the node averaged fluxes [49, 119]. Nodal methods play a pivotal role in situations where computational time is of utmost significance. Nodal methods for 'coupled reactors' were originally proposed by Avery [8]. There, 'reactors' are arbitrarily defined as subregions of the system in which fission neutrons are emitted, and the term 'coupled' means that in each of the reactors, some of the fission neutrons are emitted in fissions induced by neutrons born in other reactors. Given the 'coupling coefficients,' Avery's equations for a reactor under consideration describe the time change of the so called 'partial–fission neutron source;' representing the neutrons emanating from fissions caused by neutrons originated from other reactors. The coupling coefficients represent the fission neutrons in the reactor under consideration, produced by one neutron which originates in the adjacent reactors. However, the mechanism of the derivation of the coupled reactor kinetics equations have been based on the physically intuitive construction. Later on, a rigorous derivation of Avery's model from the time dependent adjoint equation of the partial adjoint flux has been obtained by Komata [57]. There, Komata refers the Avery's equations as a 'multipoint kinetics' model as opposed to a 'one–point kinetics model.' Many other contributions on multipoint kinetics model, for example, Difilippo and Waldman [20], Shinkawa *et.al.* [94], Kobayashi and Yoshikuni [53] and Ravetto and Rostagno [84] can be found. It has been shown by Kobayashi [53] that the static and kinetic nodal equations for the fission neutron sources in each of the coupled reactors could be derived in a more simpler and realistic manner respectively from the static and time–dependent multigroup diffusion equations, without any approximations. It has been shown that static and time–dependent coupling coefficients and neutron life time can be explicitly expressed in terms of a Green's function obtained by solving the adjoint multigroup equation of the diffusion equation. In another work [52], Kobayashi demonstrated that this approach can be extended to the multigroup transport theory, and that the coupling coefficients and the neutron life time can be expressed using angular flux and an importance function $G_m(r,\Omega,g)$ representing the number of fission neutrons produced in the region under consideration by a neutron born at position r with direction Ω and in the energy group g.

Due to several approximations made in derivations, some of which are difficult to justify, the computational approaches have to be validated numerically. A comparison of numerical results, originally published in the years 1965-69, is reported by Hetrick[38]. F. N. McDonnell *et. al.* have presented the description of some benchmark problems on CANDU reactor kinetics [65]. A benchmark problem book prepared by the Computational Benchmark Problems Committee of the Mathematics and Computational Division of the American Nuclear Society gives a collection of important problems on different reactor types [19].

1.2 Challenges in Control of Large Nuclear Reactors

Operation and control of large nuclear reactors pose several challenges. First and foremost difficulty is the development of a suitable mathematical model, which in itself is the most challenging task. Nuclear reactors of small core size are adequately represented by the well–known point kinetics model which depicts the dynamics of the core averaged neutron flux or power and the associated delayed neutron precursor groups. For large reactors, however, this model will not suffice since the flux shape in large reactors undergoes variations which the point kinetics model is unable to capture. Hence for large nuclear reactors, it is essential to adopt more detailed space–time kinetics models for analysis. However, such models often will be of very large order which hamper their direct use in controller design applications. Hence it becomes essential to develop a suitable model which captures all the essential properties of the reactor core, at the same time having a trade–off between the model accuracy and model order.

Large nuclear reactors often exhibit a property called 'loose neutronic coupling' whereby different regions of the reactor core behave differently essentially due to physical dimensions considerably greater than the neutron migration length coupled with the reactivity feedback contributed by changes in the xenon concentration. Due to this, a serious situation may arise in a large nuclear reactor in which different regions of the core may undergo variations in neutron flux in opposite phase. If the oscillations in the power distribution are left uncontrolled, the power density and the rate of change of power at some locations in the reactor core may exceed their respective thermal limits, resulting into increased chances of fuel failure. So, in large thermal nuclear reactors, it becomes necessary to employ automatic power distribution control systems or spatial control systems, besides the system for control of global power. The objective is to maintain the core power distribution close to a desired shape.

The problem of control of small nuclear reactors within the framework of point kinetic model has been extensively studied [24, 32, 38, 67]. Applications to spatial control can also be found[5, 17, 80, 110]. Application of singular perturbation approach to a different type of reactor control problem is given by Reddy and Sannuti[88]. Karppinen has given an excellent survey of spatial reactor control methods[50] employing optimization techniques. Spatial control systems employed in pressurized water reactors are discussed by Aleite[1] and those employed in CANDU reactors is discussed by Hinchley and Kuglar[40]. The effects of xenon on the kinetics of reactors has been extensively explored. Randall [83] has established an approximate criterion that can be used to find whether the reactor having certain physical dimensions would have susceptibility of xenon-induced oscillations. Canosa and Brooks [14] have further refined the criterion and explained why spatial oscillations depend strongly on the core size.

Several methods based on modern control techniques are also proposed for control of spatial power distribution in large nuclear reactors. Tiwari *et.al* [115] proposed piecewise constant periodic output feedback for spatial control of 540 MWe PHWR. Decentralized periodic output feedback technique has been adopted by Talange *et.al.* [113] for spatial control of the same reactor. Fast output sampling technique has also been successfully employed for spatial control of large nuclear reactors [10, 108, 109]. Later on, techniques based on sliding mode control [86, 87, 88] have been developed by Reddy *et.al.* for spatial control of large PHWRs.

In the subsequent chapters of this monograph, techniques for mathematical modeling and control of a large nuclear reactor have been introduced with specific applications to the Advanced Heavy Water Reactor, a thermal nuclear reactor being developed in India with the specific aim of utilizing thorium for power generation.

1.3 Advanced Heavy Water Reactor

The importance of nuclear energy as a sustainable energy resource for India was recognized at the very inception of its atomic energy programme more than four decades ago. A three-stage nuclear power programme, based on a closed nuclear fuel cycle, was then chalked out. The three stages envisaged are:

1. Natural uranium fuelled Pressurised Heavy Water Reactors,
2. Fast Breeder Reactors (FBRs) utilising plutonium based fuel, and,
3. Advanced nuclear power systems for utilisation of thorium.

Indigenous development of nuclear power plants based on uranium cycle in PHWRs was encompassed in the first stage. This enabled efficient production of plutonium, the fissile material needed to fuel further growth in nuclear power capacity. As a part of second stage, Fast Breeder Reactor programs were launched with the Fast Breeder Test Reactor (FBTR), which operates with a uranium–plutonium mixed carbide fuel. Experience from FBTR enabled the design of 500MWe Prototype Fast Breeder Reactor (PFBR) that utilizes plutonium and depleted uranium from the PHWRs.

With India's five to six times larger reserves of thorium than that of natural uranium, thorium utilization for large scale energy production has been an important goal of the third stage of the nuclear power programme. The Advanced Heavy Water Reactor (AHWR) provides a focal point for a time bound high intensity development in the efficient commercial utilization of thorium and thereby forms an important milestone in the third stage of the nuclear program [104].

The AHWR is a 920 MW(th) vertical pressure tube type reactor cooled by boiling light water and moderated by heavy water employing thorium based fuel [103]. Thorium is a fertile material and has to be converted into ^{233}U, a fissile isotope. Of the three fissile species (^{233}U, ^{235}U and ^{239}Pu), ^{233}U has the highest value of η (number of neutrons liberated for every neutron absorbed in the fuel) in thermal spectrum. Since ^{233}U does not occur in the nature, it is desirable that any system that uses ^{233}U should be self–sustaining in this nuclide in the entire fuel cycle, which implies that the amount of ^{233}U used in the cycle should be equal to the amount produced and recovered. Thorium in its natural state does not contain any fissile isotope the way uranium does. Hence, with thorium–based fuel, enrichment with fissile material is essential. Therefore the AHWR is fuelled with $(Th^{233}U)O_2$ and $(ThPu)O_2$. The fuel is designed to maximize generation of energy from thorium, to maintain self–sufficiency in ^{233}U and to achieve a slightly negative void coefficient of reactivity. The AHWR is designed to generate 300MW electrical power output.

The PHWRs employ heavy water as coolant to extract the fission heat from the fuel bundles. However, on account of its high cost and its association with radioactive tritium, use of heavy water coolant requires implementation of a costly heavy water management and recovery system. On the other hand, the large absorption cross–section for thermal neutrons in thorium facilitates the use of light water as coolant in thorium–based fuel reactors. The use of light water as coolant makes it possible to use boiling in the core, thus producing steam at a higher pressure than otherwise possible with a pressurized non–boiling system. With boiling coolant, the reactor has to be vertical, making full core heat removal by natural circulation feasible. The choice of heavy water as moderator is derived from its excellent fuel utilization characteristics. Considering these characteristics, the mainly thorium fuelled AHWR, is heavy water moderated, boiling light water cooled, and has a vertical core[104].

The reactor core of the AHWR consists of 513 lattice locations in a square lattice pitch of 225 mm, as shown in Fig. 1.1. Of these, 53 locations are for the reactivity control devices and shut down systems. Reactivity control is achieved by on–line fuelling, boron dissolved in moderator and reactivity devices. Boron in moderator is used for reactivity management of equilibrium xenon load. There are 24 control rods, grouped into regulating rods (RRs), absorber rods (ARs) and shim rods (SRs) of 8 each. The reactor has two independent, functionally diverse, fast acting shut

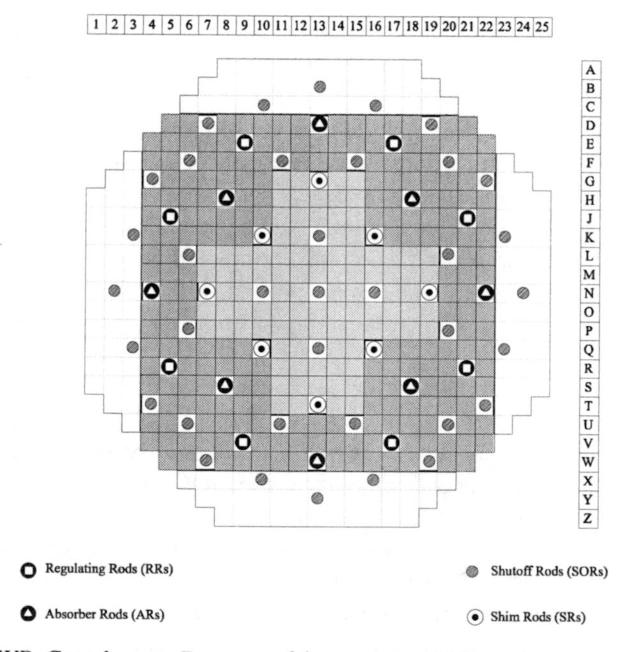

Fig. 1.1 AHWR Core layout: Burnups of innermost, middle and outermost regions are 43500, 33000 and 31500 MWD/Te respectively. Reactivity devices occupy the lattice locations, as shown.

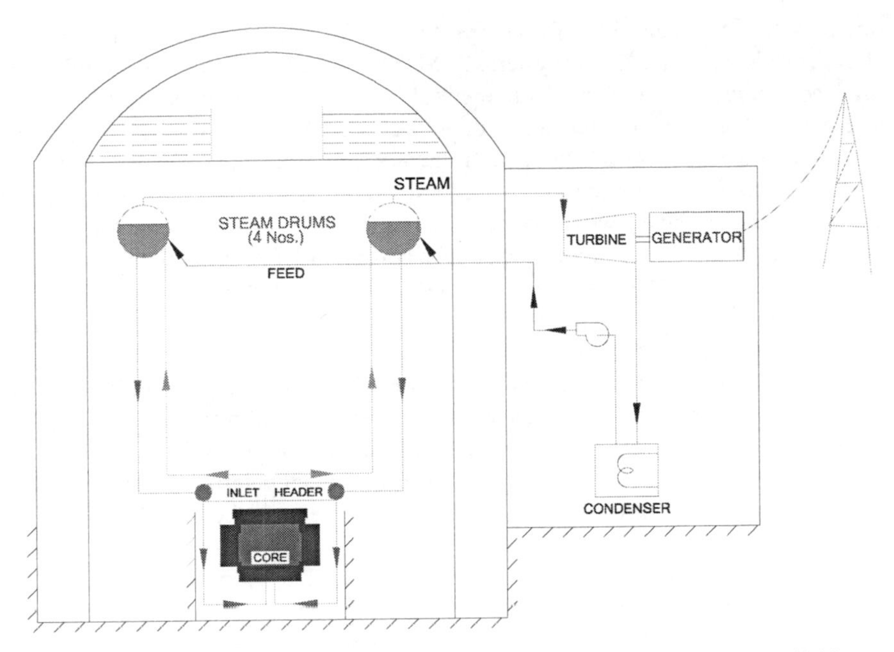

Fig. 1.2 AHWR Plant Schematic: Only two steam drums, out of the four available, are shown.

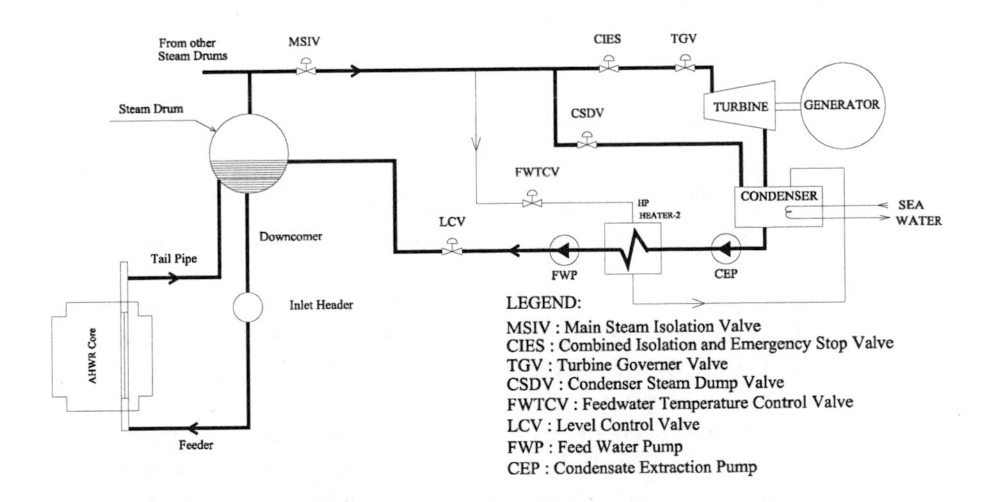

Fig. 1.3 Schematic of major secondary side components of AHWR.

down systems, namely, Shut Down System–1 (SDS–1) consisting of mechanical shut off rods and Shut Down System–2 (SDS–2) based on liquid poison injection into the moderator. There are 30 interstitial lattice locations housing 150 in–core self-powered neutron detectors and 6 out–of–core locations containing 9 ion chambers and 3 start–up detectors for measurement of neutron flux.

The reactor core is housed in a low–pressure reactor vessel called calandria. The calandria contains heavy water, which act as moderator as well as reflector. The calandria houses the vertical coolant channels, consisting of pressure tubes containing the fuel clusters. A calandria tube envelops each pressure tube and the air annulus between the two tubes provides thermal insulation between the hot coolant channel and the cold moderator.

The light water coolant picks up nuclear heat in boiling mode from fuel assemblies. The coolant circulation is driven by natural convection through tail pipes to steam drums, where steam is separated and is supplied to the turbine, as shown in Fig. 1.2. Steam separated from four steam drums, each catering to one–fourth of the core, is fed to the turbine through a set of main steam isolation valves (MSIVs) and turbine governer valves (TGVs), as shown in Fig. 1.3. The condensate, after heating using a set of heat exchangers, is pumped back to the steam drums through level control valves (LCVs) which regulate the feed flow to the steam drums so as to regulate the water level in them. The feed water temperature is maintained at 130^oC under full power operating conditions to provide optimum sub–cooling at the reactor inlet. Four down–comers, from each steam drum, are connected to a circular inlet header. The inlet header distributes the flow to each of the 452 coolant channels through individual feeders.

1.4 Overview of the Monograph

The remaining chapters of this monograph are organized as follows:

1. A mathematical model representing the coupled space–time neutronics–thermal hydraulics behavior of AHWR has been developed from first principles within the framework of nodal modeling, in Chapter 2. The number of "regions" or "nodes" in which the reactor is considered to be divided has been arrived at, in such a way that the requisite accuracy is maintained with a model of reasonable order only. Owing to the accuracy, simplicity and smaller order, this model is useful for simulation of control related transients and controller design.
2. Total and spatial power controllers for AHWR have been designed using static output feedback technique in a conventional framework, and the effect of output feedback on system stability has been brought out in Chapter 3.
3. A technique for direct decomposition of a non–autonomous three–time–scale system into three smaller order subsystems, and design of a composite controller from the three subsystem controllers is described in Chapter 4. A novel methodology for decomposition of the optimal control problem of the original high order system into three smaller order optimal control problems, with separate quadratic

performance indices extracted from the quadratic performance index of the original system, is also presented. The proposed method is applied to the spatial control of AHWR.

4. A method for direct block–diagonalization of three–time–scale systems in discrete time domain is applied on the AHWR model to obtain three decoupled smaller order subsystem models, in Chapter 5. A new strategy for derivation of composite controller from the three subsystem controllers is brought out and applied to obtain a spatial power controller for AHWR.
5. A novel fast output sampling technique has been proposed for the class of three–time–scale systems, in Chapter 6. The method eliminates the ill–conditioning effects of the existing fast output sampling approach when used for the systems with three–time–scale property. This method is also applied to the spatial control problem of AHWR.

Chapter 2
Multipoint Kinetics Modeling of Large Nuclear Reactors

2.1 Introduction

The central problem in reactor analysis is the determination of the spatial flux and power distribution in the reactor core under steady state as well as transient operating conditions. There is, however, a considerable variation in the degree of accuracy and spatial details of the power distribution required in different facets of reactor analysis and design. Basically, the behavior of neutrons in a nuclear reactor is adequately described by the time–dependent Boltzmann transport equation [24, 38]. However, numerical solutions of the coupled time–dependent transport and delayed neutron precursors' equations for reactor kinetics problems of practical interest are prohibitively difficult. So, approximate methods using the time–dependent group diffusion equations are generally employed. All of these methods, in their fundamental form, target the reconstruction of accurate space–time dependent group flux distributions over the reactor core. In many applications it is also targeted to reconstruct flux distributions up to the fuel bundle or even the fuel pin level. However, minute details of that extent are not required in control system studies, where the prime objective would usually be to model the total power and a coarse spatial power distribution within the reactor core with reasonable accuracy. In most of the control related applications, it would be preferable to cast the model equations as a set of first order differential equations so that they can be expressed in standard state space form. Among the various methods, multipoint kinetics methods, from which simple first order equations governing node averaged power can be cast rather easily, form a better candidate under such scenarios. Moreover, with transient simulations being an integral part of the control system design and analysis task, such models further establish their relevance in control related applications. Over the period of time, multipoint kinetics models have been used extensively for the analysis and simulation of light water reactors and control system design of pressurized heavy water reactors [13, 85, 86, 113, 114, 117]. In this chapter, the methodology to develop a

S.R. Shimjith et al.: Modeling and Control of a Large Nuclear Reactor, LNCIS 431, pp. 17–60.
springerlink.com

simple multipoint kinetics model with application to Advanced Heavy Water Reactor (AHWR) is described along with a control theory oriented approach for selection of suitable number of nodes.

2.2 Derivation of Multipoint Kinetics Model

The multipoint kinetics model can be derived from multigroup neutron diffusion equations, as illustrated in the following with the help of two group equation and the associated equations for delayed neutron precursors' concentrations, *i.e.*,

$$\begin{aligned} \frac{1}{v_1}\frac{\partial \phi_1}{\partial t} &= \nabla D_1 \nabla \phi_1 - \Sigma_{a1}\phi_1 - \Sigma_{12}\phi_1 + \Sigma_{21}\phi_2 \\ &\quad +(1-\beta)(\nu\Sigma_{f1}\phi_1 + \nu\Sigma_{f2}\phi_2) + \sum_{i=1}^{m_d} \lambda_i \tilde{C}_i, \end{aligned} \tag{2.1}$$

$$\frac{1}{v_2}\frac{\partial \phi_2}{\partial t} = \nabla D_2 \nabla \phi_2 - \Sigma_{a2}\phi_2 + \Sigma_{12}\phi_1 - \Sigma_{21}\phi_2, \tag{2.2}$$

$$\frac{\partial \tilde{C}_i}{\partial t} = \beta_i(\nu\Sigma_{f1}\phi_1 + \nu\Sigma_{f2}\phi_2) - \lambda_i \tilde{C}_i; \; i = 1,2,\cdots m_d, \tag{2.3}$$

where ϕ_1 and ϕ_2 denote the fast and thermal neutron fluxes respectively, m_d is the number of delayed neutron groups under consideration and $\tilde{C}_i$ is the precursor concentration corresponding to the ith delayed neutron group. It is assumed that all the fission neutrons are generated as the fast neutrons. The upscattering term $\Sigma_{21}\phi_2$ in (2.1), in general, need not be considered for thermal reactors. However, for reactors like AHWR which operate with a slightly harder spectrum in the epithermal region, the contribution of up–scattering, though small, needs to be accounted.

It should be noted that the neutron fluxes ϕ_1 and ϕ_2 are functions of both space coordinates and time. Also, the parameters $D_1, D_2, \Sigma_{a1}, \Sigma_{a2}, \Sigma_{f1}$ and Σ_{f2} are different at different core locations. Now consider dividing the reactor into a number of small coarse mesh boxes (nodes). Within each mesh box, the neutron fluxes and other neutronic parameters are represented by the respective average values integrated over its volume. Then the neutron leakage terms in (2.1) and (2.2) can be expressed approximately as

$$\nabla D_1 \nabla \phi_1 \simeq D_1 \nabla^2 \phi_{1|_h} \tag{2.4}$$

$$\nabla D_2 \nabla \phi_2 \simeq D_2 \nabla^2 \phi_{2|_h}. \tag{2.5}$$

The net rate of fast neutron flow from a box designated by 'h' to its neighbour designated by 'k' can be written as

$$D_1 \frac{d^2 \phi_1}{dr^2} V_h = J_r A_{hk}, \;\; r = x, y, \text{ or } z, \tag{2.6}$$

where V_h denotes the volume of box h and A_{hk} is the area of interface between the boxes h and k, perpendicular to the direction of flow of neutron current. Using Fick's law, the above equation can be manipulated to obtain

$$D_1 \frac{d^2\phi_1}{dr^2} = \frac{A_{hk}}{V_h} D_1 \frac{d\phi_1}{dr} = \frac{D_1 A_{hk}}{V_h \Delta_{hk}} \left[-\phi_{1_h} + \phi_{1_k}\right], \tag{2.7}$$

in which Δ_{hk} denotes the centre to centre distance between the two boxes. Based on the above, an approximation for the leakage terms in (2.4) and (2.5) could be

$$D_1 \nabla^2 \phi_1|_h = -\omega_{1hh}\phi_{1h} + \sum_{k=1}^{Z} \omega_{1hk}\phi_{1k} \tag{2.8}$$

$$D_2 \nabla^2 \phi_2|_h = -\omega_{2hh}\phi_{2h} + \sum_{k=1}^{Z} \omega_{2hk}\phi_{2k} \tag{2.9}$$

where

$$\omega_{ihk} = \frac{D_i A_{hk}}{V_h \Delta_{hk}} \text{ and } \omega_{ihh} = \sum_{k=1}^{Z} \omega_{ihk},\ i = 1,2.$$

The summations in equations (2.8) and (2.9) are not required over all mesh boxes because $\omega_{ihk} = 0$ if the kth box is not a neighbour of the hth box. Substituting the above in (2.1) and (2.2), we have

$$\begin{aligned} \frac{1}{v_{1h}} \frac{d\phi_{1h}}{dt} = &-\omega_{1hh}\phi_{1h} + \sum_{k=1}^{Z} \omega_{1hk}\phi_{1k} + (1-\beta)\left(\nu\Sigma_{f1h}\phi_{1h} + \nu\Sigma_{f2h}\phi_{2h}\right) \\ &-\Sigma_{a1h}\phi_{1h} - \Sigma_{12h}\phi_{1h} + \Sigma_{21h}\phi_{2h} + \sum_{i=1}^{m_d} \lambda_i \tilde{C}_{ih} \end{aligned} \tag{2.10}$$

$$\frac{1}{v_{2h}} \frac{d\phi_{2h}}{dt} = -\omega_{2hh}\phi_{2h} + \sum_{k=1}^{Z} \omega_{2hk}\phi_{2k} - \Sigma_{a2h}\phi_{2h} + \Sigma_{12h}\phi_{1h} - \Sigma_{21h}\phi_{2h}. \tag{2.11}$$

It can be seen that the group flux ϕ_{1h} and ϕ_{2h} depend only on time. Adding (2.10) and (2.11) and defining

$$\omega_{hh} = \frac{\omega_{1hh} + \omega_{2hh}R_h}{1+R_h},\quad \omega_{hk} = \frac{\omega_{1hk} + \omega_{2hk}R_h}{1+R_h}, \tag{2.12}$$

$$\Sigma_{ah} = \frac{\Sigma_{a1h} + \Sigma_{a2h}R_h}{1+R_h},\quad \Sigma_{fh} = \frac{\Sigma_{f1h} + \Sigma_{f2h}R_h}{1+R_h},$$

$$v_h = \frac{1+R_h}{\frac{1}{v_{1h}} + \frac{R_h}{v_{2h}}},\quad \phi_h = \phi_{1h} + \phi_{2h},\ R_h = \frac{\phi_{2h}}{\phi_{1h}}, \tag{2.13}$$

we get

$$\frac{1}{v_h}\frac{d\phi_h}{dt} = -\omega_{hh}\phi_h + \sum_{k=1}^{Z}\omega_{hk}\phi_k - \Sigma_{ah}\phi_h + (1-\beta)\nu\Sigma_{fh}\phi_h + \sum_{i=1}^{m_d}\lambda_i\tilde{C}_{ih}. \quad (2.14)$$

From (2.13) it is evident that for computation of different coefficients and parameters in equivalent one group equation (2.14), knowledge of thermal–to–fast flux ratio for each mesh box is essential. A straightforward approach would be to solve (2.10) and (2.11) at steady state for a representative core configuration and then compute R_h for each mesh box in the core. However, this would be computationally expensive. In an alternate approach, steady state form of (2.11) is considered and leakage of thermal neutrons is ignored to obtain

$$0 = -\Sigma_{a2h}\phi_{2h} + \Sigma_{12h}\phi_{1h} - \Sigma_{21h}\phi_{2h}$$

from which it follows that

$$R_h = \frac{\phi_{1h}}{\phi_{2h}} \approx \frac{\Sigma_{12h}}{\Sigma_{a2h} + \Sigma_{21h}}. \quad (2.15)$$

Also, from (2.3), variation of delayed neutron precursors' concentration in box h can be described by

$$\frac{d\tilde{C}_{ih}}{dt} = \beta_i\nu\Sigma_{fh}\phi_h - \lambda_i\tilde{C}_{ih}; \;\; i = 1,2,\cdots m_d. \quad (2.16)$$

The one group equation (2.14) alongwith the delayed neutron precursors' equation (2.16) can be used to study the nuclear reactor. However, we proceed further by expressing (2.14) in an alternative form more suitable for control studies. Let E_{eff} be the average thermal energy liberated in each fission. Then the mesh box power is obtained from neutron flux as

$$Q_h = E_{eff}\Sigma_{fh}V_h\phi_h. \quad (2.17)$$

Likewise the delayed neutron precursors' concentrations in mesh box h can be modified as

$$C_{ih} = \tilde{C}_{ih}E_{eff}\Sigma_{fh}V_h v_h. \quad (2.18)$$

Using these new variables we can obtain respectively from (2.14) and (2.16):

$$\begin{aligned}\frac{dQ_h}{dt} &= -\omega_{hh}v_hQ_h + \sum_{k=1}^{Z}\frac{\omega_{hk}\Sigma_{fh}V_h}{\Sigma_{fk}V_k}v_hQ_k - \Sigma_{ah}v_hQ_h \\ &\quad +(1-\beta)\nu\Sigma_{fh}v_hQ_h + \sum_{i=1}^{m_d}\lambda_iC_{ih}, \quad (2.19)\\ \frac{dC_{ih}}{dt} &= \beta_{ih}\nu\Sigma_{fh}v_hQ_h - \lambda_iC_{ih}, \;\; i = 1,2,\cdots m_d. \quad (2.20)\end{aligned}$$

Substituting the value of R_h from (2.15) in the expression for ω_{hk}, we get

$$\omega_{hk} = D_h \frac{A_{hk}}{V_h \Delta_{hk}},$$
$$\text{where, } D_h = \frac{(\Sigma_{a2h} + \Sigma_{21h}) D_1 + \Sigma_{12h} D_2}{\Sigma_{a2h} + \Sigma_{12h} + \Sigma_{21h}},$$
$$\text{by which, } \omega_{hk} \frac{V_h}{V_k} = D_h \frac{A_{hk}}{V_k \Delta_{hk}} = \omega_{kh}.$$

Further by defining the prompt neutron lifetime and infinite multiplication factors for the mesh boxes/nodes as $\ell_h = \frac{1}{\Sigma_{ah} v_h}$ and $K_h = \frac{\nu \Sigma_{fh}}{\Sigma_{ah}}$, reactivity in node h as $\rho_h = \frac{K_h - 1}{K_h}$, and

$$\alpha_{hh} = \omega_{hh} v_h \ell_h \text{ and } \alpha_{kh} = \frac{\omega_{hk} \Sigma_{fh} V_h v_h \ell_h}{\Sigma_{fk} V_k},$$

we get

$$\frac{dQ_h}{dt} = -\alpha_{hh} \frac{Q_h}{\ell_h} + \sum_{k=1}^{Z} \alpha_{kh} \frac{Q_k}{\ell_h} + (\rho_h - \beta) \frac{Q_h}{\ell_h} + \sum_{i=1}^{m_d} \lambda_i C_{ih}, \tag{2.21}$$

$$\frac{dC_{ih}}{dt} = \frac{\beta_i}{\ell_h} Q_h - \lambda_i C_{ih}, \quad h = 1, 2, \cdots Z, \tag{2.22}$$

where α_{kh} and α_{hh}, which are functions of ω_{hk} and ω_{hh} respectively, denote the coupling between kth and hth nodes and the self coupling coefficient of hth node. Equations (2.21) and (2.22) represent the neutronics model of the reactor core without internal feedbacks, *e.g.*, the fission product reactivity feedback.

The power generated in a reflector mesh box is zero. Hence, description of the core by (2.21) and (2.22) requires fewer equations than (2.14) does. However, the neutron leakage to reflector mesh boxes needs to be accounted. For this, the neutron leakage between node h in the core region and the neighboring non-power generating reflector nodes can be clubbed with ω_{hh} of node h.

In situations where the reference steady state power distribution $\hat{Q}$ is available, ω_{hh} for each node can alternatively be calculated using the reference power distribution itself while the cross–leakage coefficients ω_{hk} are still calculated using (2.12). From the steady state forms of (2.19) and (2.20), we get

$$\omega_{hh} = \frac{\nu \Sigma_{fh} \hat{Q}_h + \sum_{k=1}^{Z} \frac{\omega_{hk} \Sigma_{fh} V_h}{\Sigma_{fk} V_k} \hat{Q}_k - \Sigma_{ah} \hat{Q}_h}{\hat{Q}_h}.$$

Now, in order to account for reactivity variations due to internal feedbacks and control devices, the reactivity term ρ_h in (2.21) is expressed as $\rho_h = \rho_{hX} + \rho_{h_u}$ where ρ_{h_u} is the reactivity introduced by the control rods and ρ_{hX} is the reactivity feedback due to xenon.

To formulate xenon reactivity feedback, iodine and xenon dynamics in each node can be modeled as:

$$\frac{dI_h}{dt} = \gamma_I \Sigma_{fh} Q_h - \lambda_I I_h, \tag{2.23}$$

$$\frac{dX_h}{dt} = \gamma_x \Sigma_{fh} Q_h + \lambda_I I_h - (\lambda_x + \overline{\sigma}_{xh} Q_h) X_h \tag{2.24}$$

where I_h and X_h are respectively the iodine and xenon concentrations in node h, γ_I and γ_x are their respective fractional yields, λ_I and λ_x are respective decay constants, and $\overline{\sigma}_{xh} = \frac{\sigma_{ax}}{E_{eff} \Sigma_{fh} V_h}$. The xenon reactivity feedback is given by:

$$\rho_{hX} = -\frac{\overline{\sigma}_{xh} X_h}{\Sigma_{ah}}.$$

The reactivity ρ_{h_u} introduced by the control rods (or in general reactivity control devices) can be expressed as a function of the instantaneous position of the device as

$$\rho_{h_u} = f(H_j);$$

where H_j is the position of the control rod and f is a function which could be readily obtained by fitting a polynomial to the calibration data of the device depicting its reactivity worth versus position. However, in this process, to make the representation complete, we need to model the dynamics of the control rod as well. In many applications, a simple model can be considered by neglecting the dynamics of the control rod drive mechanism, friction, damping and rotational to linear motion transmission dynamics, with which the speed of control rod is assumed directly proportional to the applied voltage v to the drive motor, *i.e.*,

$$\frac{dH_j}{dt} = K v_j. \tag{2.25}$$

The set of equations (2.21) - (2.25) describe the variation of power levels and delayed neutron precursors' concentrations in different nodes of the reactor, alongwith the variation of iodine and xenon concentrations in each node and the dynamics of the regulating rod drives. This set of equations represents a very simple model of AHWR for study of neutronics behavior of the core. This model can be augmented with the simplified model of core thermal hydraulics to make the representation complete.

Remarks

Coupling coefficients and other parameters in (2.21) have been derived in terms of parameters of one group neutron diffusion equation (2.14). These one group parameters in turn have been obtained from two group parameters appearing in (2.10) and

(2.11). This approach, though simple, yields satisfactory accuracy. It may, however, also be possible to obtain the coupling coefficients and other parameters in (2.21) directly in terms of the two group parameters. Moreover, if the two group fluxes are known then the assumption of the spectral ratio given by (2.15) would not be required.

Instead of (2.21)–(2.25), the set of equations (2.10) and (2.11), alongwith the associated equations for delayed neutron precursors', iodine and xenon concentrations and the control rod equation can also be considered. This might improve the accuracy by eliminating the approximations in group collapsing, however, at the cost of increase in model order.

2.2.1 *Linearization of Model Equations*

The set of first order differential equations (2.21) - (2.25) can easily be cast in standard state space form amenable for control studies. Let Q_{h_0} denote the fission power in node h, corresponding to the full power operation of the reactor. With this, corresponding equilibrium values of C_{ih_0}, I_{h_0} and X_{h_0} denoting the delayed neutron precursors, iodine and xenon concentrations respectively can readily be computed from the steady state forms of the respective equations (2.22), (2.23) and (2.24). Also let H_{j_0} denote the equilibrium position of jth control rod. The set of equations (2.21), (2.22), (2.23), (2.24) and (2.25) can be linearized around these steady state conditions. Define the state and input vectors as

$$z = \begin{bmatrix} \delta Q \ \delta C \ \delta I \ \delta X \ \delta H \end{bmatrix}, \tag{2.26}$$

$$u = \begin{bmatrix} \delta v_2 \ \delta v_4 \ \delta v_6 \ \delta v_8 \end{bmatrix}^T, \tag{2.27}$$

then the equations (2.21) – (2.25) can be readily arranged in a matrix form, leading to a linear state space model

$$\begin{aligned} \dot{z} &= Az + Bu \\ y &= Mz. \end{aligned} \tag{2.28}$$

The system matrix A is of order $Z(m_d+3)+N_{RR}$, where Z is the number of nodes under consideration and N_{RR} is the number of control rods available for automatic control. A is expressed as

$$A = \begin{bmatrix} A_{QQ} & A_{QC} & 0 & A_{QX} & A_{QH} \\ A_{CQ} & A_{CC} & 0 & 0 & 0 \\ A_{IQ} & 0 & A_{II} & 0 & 0 \\ A_{XQ} & 0 & A_{XI} & A_{XX} & 0 \\ 0 & 0 & 0 & 0 & 0 \end{bmatrix} \tag{2.29}$$

where the last row corresponds to N_{RR} rows of zeros, and 0 stands for null matrix of appropriate dimensions. Remaining sub blocks are given by:

$$A_{QQ}(i,j) = \begin{cases} \frac{1}{\ell}\left(-\sum_{k=1}^{Z} \alpha_{ki}\frac{Q_{k_0}}{Q_{i_0}} - \beta\right) & \text{if } i=j \\ \frac{1}{\ell}\alpha_{ji}.\frac{Q_{j_0}}{Q_{i_0}} & \text{if } i \neq j \end{cases}$$

$$A_{QC} = diag.\left[a_{qc1}\ \ a_{qc2}\cdots a_{qcZ}\right];$$

$$\text{where } a_{qci} = \frac{1}{\ell_i}\left[\beta_1 \cdots \beta_{m_d}\right],$$

$$A_{QX} = -\frac{1}{\ell}diag.\left[\frac{\overline{\sigma}_{x1}}{\Sigma_{a1}}X_{1_0}\ \ \frac{\overline{\sigma}_{x2}}{\Sigma_{a2}}X_{2_0}\cdots\frac{\overline{\sigma}_{xZ}}{\Sigma_{aZ}}X_{Z_0}\right],$$

$$A_{CQ} = diag.\left[\lambda_1\ \ \lambda_2\cdots\lambda_Z\right];$$

$$\text{where } \lambda_i = \left[\lambda_1 \cdots \lambda_{m_d}\right]^T,$$

$$A_{CC} = diag.\left[\overline{\lambda}_1\ \ \overline{\lambda}_2\cdots\overline{\lambda}_Z\right];$$

$$\text{where } \overline{\lambda}_i = diag.\left[\lambda_1 \ \cdots\ \lambda_Z\right],$$

$$A_{IQ} = \lambda_I E_Z \text{ and } A_{II} = -A_{IQ},$$

$$A_{XQ} = diag.\left[\lambda_x - \lambda_I\frac{I_{1_0}}{X_{1_0}}\ \ \lambda_x - \lambda_I\frac{I_{2_0}}{X_{2_0}}\cdots\lambda_x - \lambda_I\frac{I_{Z_0}}{X_{Z_0}}\right],$$

$$A_{XX} = -diag.\left[\lambda_x + \overline{\sigma}_{x1}Q_{1_0}\ \ \lambda_x + \overline{\sigma}_{x2}Q_{2_0}\cdots\lambda_x + \overline{\sigma}_{xZ}Q_{Z_0}\right],$$

$$A_{XI} = \lambda_I diag.\left[\frac{I_{1_0}}{X_{1_0}}\ \ \frac{I_{2_0}}{X_{2_0}}\cdots\frac{I_{Z_0}}{X_{Z_0}}\right],$$

$$A_{QH}(i,j) = \begin{cases} \frac{Q_{i_0}}{\ell}\frac{\partial f(H_j)}{\partial H_j} & \text{if node } i \text{ contains the control rod } j, \\ 0 & \text{elsewhere} \end{cases}.$$

Matrix B is of dimension $(Z(m_d+3)+N_{RR}) \times N_{RR}$, as follows:

$$B = \left[0\ 0\ 0\ 0\ B_H^T\right]^T; \tag{2.30}$$

where B_H is a diagonal matrix of dimension $N_{RR} \times N_{RR}$, with K as diagonal entries.

The total power of the reactor is monitored using out–of–core ion chamber and the spatial flux distribution within the core is monitored by in–core detectors. Output of the ion chamber could be modelled as $\kappa_1 \sum_{h=1}^{Z} Q_h$, where κ_1 is a constant depending on the ion chamber sensitivity and gain of amplifier. Likewise, the output of an incore detector can be modelled as $\kappa_2 \sum \psi_{ij} Q_j$, where $\psi_{ij} = 1$ if the ith detector is near j^{th} zone, and ψ_{ij} is zero otherwise. κ_2 is another parameter depending upon the sensitivity of the in–core detector and the gain of the amplifier. Assuming m as the total number of detectors available for power measurement,

$$M = [M_1\ \ M_2], \tag{2.31}$$

is of dimension $m \times [Z(m_d+3)+N_{RR}]$, where M_1 and M_2 are given by

$$M_1 = \begin{bmatrix} \kappa_1 & \kappa_1 & \cdots & \kappa_1 \\ \kappa_2 & 0 & \cdots & 0 \\ 0 & \kappa_2 & \cdots & 0 \\ \vdots & \vdots & \ddots & \vdots \\ 0 & 0 & \cdots & \kappa_2 \end{bmatrix}, \qquad (2.32)$$

$$M_2 = 0. \qquad (2.33)$$

The first row of M_1 corresponds to the total power output sensed by the ion chambers (excore), while remaining rows correspond to zonal power outputs sensed by the incore detectors, each of which is assumed to be influenced by the power level in the zone in which it is located.

2.3 Selection of Suitable Nodalization Scheme

It is evident that the model order, denoted by the number of first order differential equations describing the neutronics phenomena in the reactor, is $Z(m_d+3)+N_{RR}$. For control system studies, a model of small order with satisfactory steady state and transient response accuracies is preferred. In the absence of benchmark solutions, the transient response of an approximate model can be compared with the response of a rigorous model to the same transient. However, it might be cumbersome to perform transient response simulation of a large set of nodalization schemes to arrive at a suitable scheme. Instead, it would be attractive if a suitable scheme can be selected based on some other model properties which are simpler to analyze. The scheme thus selected must further be qualified for its transient response accuracy.

It is generally expected that a scheme with a very large number of fine meshes or nodes will reproduce the power distribution of the reactor core with good accuracy, and at the same time it will also capture all the essential properties of the reactor, but the model order will be correspondingly very large. From the linear control theoretic point of view, a lower order model would be acceptable if stability, controllability and observability properties depicted by it are identical to those depicted by a model with a very large number of nodes. In other words, a nodalization scheme yielding a satisfactory steady state accuracy with the minimum number of nodes, yet able to reproduce the linear system properties of a very high order model, can be considered. These essential properties are introduced in the following.

2.3.1 *Steady State Accuracy*

For each nodalization scheme under consideration, the steady state core power distribution can be computed from the steady state form of nodal model equations. At steady state, the delayed neutron precursors' concentration and iodine and xenon concentrations are in equilibrium with the power. Also the control rods are at positions corresponding to the critical core configuration. In such a situation, we obtain from (2.19) and (2.20),

$$(\Sigma_{ah} + \omega_{hh}) Q_h - \sum_{k=1}^{Z} \frac{\omega_{hk}\Sigma_{fh}V_h}{\Sigma_{fk}V_k} Q_k = \nu\Sigma_{fh}Q_h. \tag{2.34}$$

Now the usual iterative criticality search algorithm [75, 107] can be used to obtain the steady state power distribution. However, the steady state power distribution can also be obtained by formulating (2.19) in the form

$$Ws = \left(\frac{1}{k_{eff}}\right) s \tag{2.35}$$

and solving for the eigenvector s corresponding to the largest value of k_{eff}, the effective multiplication factor, where

$$W = \begin{bmatrix} \overline{\omega}_{11} & -\overline{\omega}_{12} & \cdots & -\overline{\omega}_{1Z} \\ -\overline{\omega}_{21} & \overline{\omega}_{22} & \cdots & -\overline{\omega}_{2Z} \\ \vdots & \vdots & \ddots & \vdots \\ -\overline{\omega}_{Z1} & -\overline{\omega}_{Z2} & \cdots & \overline{\omega}_{ZZ} \end{bmatrix},$$

$$\overline{\omega}_{hh} = \frac{\Sigma_{ah} + \omega_{hh}}{\nu\Sigma_{fh}}, \quad \overline{\omega}_{hk} = \frac{\omega_{hk}\Sigma_{fh}V_h}{\nu\Sigma_{fk}\Sigma_{fk}V_k},$$

and $s(h) = \nu\Sigma_{fh}Q_h$.

The nodal powers thus computed can be compared with an available reference power distribution for assessing the steady state accuracy. Further, assuming that the power distribution within a node is uniform, a steady state channel power distribution can also be obtained from the nodal model. Having obtained this, comparison of different schemes in terms of their accuracy in computation of channel power distribution can also be made.

2.3.2 Linear System Properties

2.3.2.1 Stability

The linear system (2.75) is said to be asymptotically stable if $Re(\varphi_i(A)) < 0 \ \forall i$, where φ_i is an eigenvalue of A. Unstable modes of the system are those φ_i's such that $Re(\varphi_i(A)) \geq 0$ [16].

Hence, stability of different schemes can be assessed by checking the eigenvalues of the corresponding open loop linear system matrix A defined by (2.29). Stability is a very important characteristic of the system, which implies that small changes in the system input, or any non–equilibrium initial conditions, do not result in large changes in system outputs. It is known that a nuclear reactor, in the absence of any reactivity feedback effects, is not asymptotically stable due to the presence of an eigenvalue at the origin. Internal reactivity feedbacks due to xenon and temperature variations further affect the stability of the system. The number of eigenvalues with non-negative real parts corresponds to the number of unstable modes of oscillations of the spatial flux (power) distribution.

2.3.2.2 Controllability

The linear system (2.75) with order n is said to be controllable if and only if $rank([\varphi_i E - A \;\; B]) = n \;\; \forall i$, where φ_i is an eigenvalue of A and E is an identity matrix of dimension n. If this is satisfied, then it is commonly mentioned that (A,B) pair is controllable. Uncontrollable modes of the system are those φ_i's for which $rank([\varphi_i E - A \;\; B]) < n$ [37].

The condition of controllability governs the existence of a complete solution to the control system design problem. Designing a controller to stabilize an unstable system and to achieve any specified transient response characteristics may not be possible if the system is uncontrollable. Although most physical systems are controllable, their simplified mathematical models may not possess this property.

2.3.2.3 Observability

The linear system (2.75) of order n is said to be observable if and only if
$rank([\varphi_i E - A \mid M]) = n \;\; \forall i.$

If this is satisfied, then it is commonly mentioned that (A,M) pair is observable. Unobservable modes of the system are those φ_i's for which $rank([\varphi_i E - A \mid M]) < n$ [37].

Observability denotes whether every state can be determined from the observation of the outputs over a finite interval of time. The concept of observability thus helps in solving the problem of reconstructing unmeasured state variables from the measured variables. This plays a significant role in control system design since the information of all the state variables are many a times essential for designing a suitable controller.

2.3.3 Method for Selection of Nodalization Scheme

For the selection of a suitable number of nodes, the following methodology can be adopted. Initially the reactor core is considered to be divided into a few nodes. To begin with, one may start with two nodes and proceed by increasing the number of nodes through subdividing each node. In each case, the steady state power distribution, calculated using (2.35), can be compared with the reference power distribution available from a detailed finite difference model. As the number of nodes are increased, degree of accuracy is expected to improve. Also, the eigenvalues of the matrix A are obtained to assess the stability of each scheme. As the number of nodes is increased, the number of eigenvalues indicating instability might change but not beyond the number indicating the overall characteristics of the core. Hence a suitable number of nodes from the stability point of view would be the one which on further subdivision does not show any change in the number of unstable modes. In other words, it is the minimum number of nodes with which the model exhibits identical unstable modes to that of a model with a very large number of nodes. Likewise, controllability and observability properties of each scheme are also assessed. Again, like stability, it is expected that beyond a certain minimum number of nodes required to adequately represent the reactor core, these properties would remain consistent. Having obtained a suitable nodalization scheme in terms of steady state accuracy and the linear system properties, it should further be ensured that it exhibits satisfactory transient response accuracy.

2.4 Application to the AHWR

The above described method of multipoint kinetics modeling and selection of nodalization scheme is now applied to the AHWR [98]. For small–scale transients involving normal operational and control situations, reactivity control requirements of AHWR are met with by regulating rods (RRs), *i.e.*, ρ_{h_u} is essentially on account of RR movements. Reactivity contributed by the movement of a RR is a non–linear function of its position. However, around the equilibrium position, the nonlinearity is very insignificant. Thus, if node h contains the RR j,

$$\rho_{h_u} \cong (-10.234H_j + 676.203) \times 10^{-6};$$

where H_j is the '% in' position of RR j. ρ_{h_u} is zero for the nodes not containing the RR.

The RRs of AHWR are grouped into two banks, each of 4 rods. RRs in node 2, 4, 6 and 8 are in first bank, and those in 3, 5, 7 and 9 form the second bank. Under normal operation, first bank is used for automatic control, whereas the other bank is kept under manual control. This design restriction is imposed to meet the constraints on maximum rate of reactivity insertion. RRs in a bank can be moved simultaneously to control the bulk reactor power, as well as individually to control the spatial power distribution. Further, in (2.25), v_j is the control signal in the range of $\pm 5V$ applied to the jth RR drive. K is a constant decided by the constraint in maximum speed of movement of regulating rods (thereby the maximum rate of reactivity insertion) under the maximum control signal, as $K = 0.56$.

The reference steady state power distribution was generated using a finite difference model similar to that described in [75]. In the finite difference model, the reactor core is considered to be divided into 12312 meshes, with 24 meshes along the reactor axis for each of the 513 lattice locations (channels). Furthermore, side reflector region was assumed to be divided into 6048 non–power generating meshes.

Table 2.1 Two Group Cross–Section Data for Different Reactor Elements (All cross–sections (Σ) in cm^{-1} and diffusion coefficients (D) in cm).

Coolant Density 0.74g/cc	Coolant Density 0.45g/cc
Fuel (43500 MWD/Te Burnup Region)	
$\Sigma_{a1} = 0.35 \times 10^{-2}$, $\Sigma_{a2} = 0.12 \times 10^{-1}$	$\Sigma_{a1} = 0.333 \times 10^{-2}$, $\Sigma_{a2} = 0.118 \times 10^{-1}$
$\Sigma_{f1} = 0.803 \times 10^{-3}$, $\Sigma_{f2} = 0.475 \times 10^{-2}$	$\Sigma_{f1} = 0.815 \times 10^{-3}$, $\Sigma_{f2} = 0.481 \times 10^{-2}$
$\nu\Sigma_{f1} = 0.209 \times 10^{-2}$, $\nu\Sigma_{f2} = 0.124 \times 10^{-1}$	$\nu\Sigma_{f1} = 0.215 \times 10^{-2}$, $\nu\Sigma_{f2} = 0.126 \times 10^{-1}$
$D_1 = 1.55$, $D_2 = 0.895$	$D_1 = 1.55$, $D_2 = 0.896$
$\Sigma_{12} = 0.792 \times 10^{-2}$, $\Sigma_{21} = 0.218 \times 10^{-3}$	$\Sigma_{12} = 0.732 \times 10^{-2}$, $\Sigma_{21} = 0.221 \times 10^{-3}$
Fuel (33000 MWD/Te Burnup Region)	
$\Sigma_{a1} = 0.356 \times 10^{-2}$, $\Sigma_{a2} = 0.124 \times 10^{-1}$	$\Sigma_{a1} = 0.332 \times 10^{-2}$, $\Sigma_{a2} = 0.118 \times 10^{-1}$
$\Sigma_{f1} = 0.846 \times 10^{-3}$, $\Sigma_{f2} = 0.508 \times 10^{-2}$	$\Sigma_{f1} = 0.814 \times 10^{-3}$, $\Sigma_{f2} = 0.480 \times 10^{-2}$
$\nu\Sigma_{f1} = 0.221 \times 10^{-2}$, $\nu\Sigma_{f2} = 0.133 \times 10^{-1}$	$\nu\Sigma_{f1} = 0.213 \times 10^{-2}$, $\nu\Sigma_{f2} = 0.125 \times 10^{-1}$
$D_1 = 1.55$, $D_2 = 0.897$	$D_1 = 1.55$, $D_2 = 0.896$
$\Sigma_{12} = 0.787 \times 10^{-2}$, $\Sigma_{21} = 0.224 \times 10^{-3}$	$\Sigma_{12} = 0.729 \times 10^{-2}$, $\Sigma_{21} = 0.220 \times 10^{-3}$
Fuel (31500 MWD/Te Burnup Region)	
$\Sigma_{a1} = 0.357 \times 10^{-2}$, $\Sigma_{a2} = 0.126 \times 10^{-1}$	$\Sigma_{a1} = 0.330 \times 10^{-2}$, $\Sigma_{a2} = 0.116 \times 10^{-1}$
$\Sigma_{f1} = 0.863 \times 10^{-3}$, $\Sigma_{f2} = 0.522 \times 10^{-2}$	$\Sigma_{f1} = 0.802 \times 10^{-3}$, $\Sigma_{f2} = 0.474 \times 10^{-2}$
$\nu\Sigma_{f1} = 0.225 \times 10^{-2}$, $\nu\Sigma_{f2} = 0.137 \times 10^{-1}$	$\nu\Sigma_{f1} = 0.202 \times 10^{-2}$, $\nu\Sigma_{f2} = 0.121 \times 10^{-1}$
$D_1 = 1.55$, $D_2 = 0.897$	$D_1 = 1.55$, $D_2 = 0.896$
$\Sigma_{12} = 0.785 \times 10^{-2}$, $\Sigma_{21} = 0.227 \times 10^{-3}$	$\Sigma_{12} = 0.728 \times 10^{-2}$, $\Sigma_{21} = 0.219 \times 10^{-3}$
RR and AR Material (for both 0.74 and 0.45 g/cc)	
$\Sigma_{a1} = 2.92189 \times 10^{-3}$, $\Sigma_{a2} = 7.25461 \times 10^{-3}$	
$D_1 = 1.2994$, $D_2 = 0.8514$	
$\Sigma_{12} = 9.01124 \times 10^{-3}$, $\Sigma_{21} = 9.91503 \times 10^{-5}$	
SR and SOR Material (for both 0.74 and 0.45 g/cc)	
$\Sigma_{a1} = 6.0723 \times 10^{-3}$, $\Sigma_{a2} = 1.9909 \times 10^{-2}$	
$D_1 = 1.27414$, $D_2 = 0.85382$	
$\Sigma_{12} = 6.62071 \times 10^{-3}$, $\Sigma_{21} = 2.57552 \times 10^{-4}$	
Reflector (for both 0.74 and 0.45 g/cc)	
$\Sigma_{a1} = 9.04666 \times 10^{-6}$, $\Sigma_{a2} = 6.47829 \times 10^{-5}$	
$D_1 = 1.3259$, $D_2 = 0.8374$	
$\Sigma_{12} = 1.14266 \times 10^{-2}$, $\Sigma_{21} = 9.26408 \times 10^{-7}$	

In addition, top and bottom reflector regions were also divided into 2295 meshes each. Fig. 2.1 shows the reference channel power distribution in a quadrant of the reactor core thereby computed under normal full power operating conditions. Reactor exhibits a quadrant core symmetry in power distribution. For arriving at a suitable nodalization scheme, in the first step, the reactor core is considered to be divided into two and three nodes as shown in Fig. 2.2 and Fig. 2.3 respectively while side reflector was divided into 8 nodes. The nodes in the core region are progressively subdivided to obtain 4, 5, 8, 9, 16, 17, 32, 33, 64 and 128 node schemes as shown in Fig. 2.4 to Fig. 2.13 respectively. In all these schemes, the side reflector is assumed to be divided identically into 8 nodes, and the top and bottom reflectors are divided in a pattern exactly identical to the core nodalization scheme.

The physical parameters such as volume of the nodes, area of interface, distance between the nodes *etc.*, and the homogeneous neutron cross-sections for the nodes under consideration are essential for computation of coupling coefficients and determination of system matrices. Study of any given nodalization scheme can proceed therefrom. However, the cross-section values vary from node to node in AHWR due to the non-uniformity of the reactor core in terms of fuel enrichment, burnup regions and coolant density. Under full power operating conditions, coolant density varies gradually from 0.98 g/cc at the core inlet to 0.2 g/cc at the core outlet, and the neutron cross–sections also vary accordingly. Tabulating the two group cross–sections to account for such a wide variation in coolant density would be enormous. However, for brevity, the average coolant densities in the bottom half (for a length of 1.75 meter from the core bottom) and top half (remaining 1.75 meter length of the channels) of the reactor core under full power operating conditions are calculated as 0.74 g/cc and 0.45 g/cc respectively, and corresponding cross–sections, listed in Table 2.1, are used in the analysis. Using the two group flux data generated from the finite–difference model, approximate two group cross–sections for each node are computed through volume–flux weighted homogenization [24] for each nodalization scheme as:

$$\Sigma_{\zeta ih} = \frac{\sum_{j\in V_h} \Sigma_{\zeta ij}\phi_{ij}V_j}{\sum_{j\in V_h} \phi_{ij}V_j}, \quad i = 1,2; \quad h = 1\cdots Z,$$

where ζ denotes the respective neutron interaction.

The steady state accuracy and stability, controllability and observability properties of different nodalization schemes are elaborated in the following subsections.

2.4.1 Steady State Accuracy

Let $\overline{Q}_j, \; j = 1,\cdots,513$, denote the steady state channel power distribution obtained from the finite difference model. Taking these as the reference, steady state reference nodal powers are obtained by summing up the powers of the channels belonging to the node for the nodalization scheme under consideration, *i.e.*, the reference nodal power

$$\hat{Q}_h = \sum_{j\in h} \overline{Q}_j, \quad h = 1\cdots Z \tag{2.36}$$

where Z is the number of nodes in the scheme, which is 2 for the scheme shown in Fig. 2.2, 3 for that in Fig. 2.3, and so on. The RMS deviations (RMSD) between the reference nodal powers and the nodal powers determined from (2.35) is

$$RMSD = \sqrt{\frac{\sum_{i=1}^{Z} \left(\hat{Q}_i - Q_i\right)^2}{Z}}$$

where Q_i are the nodal powers obtained from (2.35).

The steady state nodal power distribution computed for each nodalization scheme is found to closely match the corresponding reference distribution, as evident from Table 2.2. Moreover, steady state accuracies in channel power distributions show a systematic improvement with the increase in number of nodes for schemes with odd number of nodes and those with even number of nodes separately.

Table 2.2 Error in Steady State Power Distribution for Different Schemes.

Scheme	RMSD (MW thermal) in computation of	
	Nodal Powers	Channel Powers
2 Node Scheme	1.048×10^{-13}	0.1895
3 Node Scheme	6.166×10^{-10}	0.1736
4 Node Scheme	4.621×10^{-12}	0.1896
5 Node Scheme	1.927×10^{-9}	0.1735
8 Node Scheme	5.274×10^{-8}	0.1896
9 Node Scheme	1.092×10^{-9}	0.1729
16 Node Scheme	3.114×10^{-9}	0.1479
17 Node Scheme	7.621×10^{-10}	0.1539
32 Node Scheme	1.935×10^{-9}	0.1242
33 Node Scheme	1.475×10^{-9}	0.1323
64 Node Scheme	4.488×10^{-9}	0.1151
128 Node Scheme	3.127×10^{-10}	0.1102

2.4.2 *Stability*

Stability of each of the above models was assessed by checking the eigenvalues of the corresponding open loop linear system matrix A defined by (2.29). An effective single group of delayed neutrons was considered for formulating the linear model, with neutronic parameters as listed in Table 2.3.

Table 2.3 Neutronic Data used in Linear Models.

β	2.643×10^{-3}
λ	$6.4568 \times 10^{-2}\ s^{-1}$
ℓ	$3.6694 \times 10^{-4}\ s$
λ_I	$2.878 \times 10^{-5}\ s^{-1}$
λ_X	$2.1 \times 10^{-5}\ s^{-1}$
γ_I	5.7×10^{-2}
γ_X	1.1×10^{-2}
σ_{aX}	$1.8 \times 10^{-22}\ cm^{-1}$
E_{eff}	$3.2 \times 10^{-11}\ J$

Each model exhibited four eigenvalues at origin when xenon reactivity feedback is considered. If the xenon reactivity feedback is not considered, each model exhibited a fifth eigenvalue at origin and all the remaining eigenvalues had negative real parts. However, when xenon reactivity feedback is considered, all the models are found to possess many eigenvalues with positive real parts. Table 2.4 lists the unstable eigenvalues of the matrix A for the different schemes, apart from the four at origin.

Table 2.4 Unstable Eigenvalues Exhibited by Different Schemes (In each scheme, there are also four eigenvalues at origin not listed here).

2 Node Scheme	3 Node Scheme	4 Node Scheme
3.4469×10^{-6}	3.4440×10^{-6}	3.4469×10^{-6}
2.8627×10^{-5}	1.8024×10^{-5}	2.8519×10^{-5}
3.0661×10^{-4}	5.2433×10^{-4}	2.8519×10^{-5}
1.0829×10^{-2}	1.0835×10^{-2}	$(1.8624 \pm j8.6157) \times 10^{-5}$
		3.0791×10^{-5}
		3.0791×10^{-4}
		1.0829×10^{-2}

5 Node Scheme	8 Node Scheme	9 Node Scheme
3.4440×10^{-6}	3.4466×10^{-6}	3.4434×10^{-6}
4.5330×10^{-5}	3.4270×10^{-5}	4.1763×10^{-5}
4.5330×10^{-5}	3.9737×10^{-5}	5.0685×10^{-5}
$(2.2419 \pm j8.5239) \times 10^{-5}$	2.1504×10^{-4}	1.6383×10^{-4}
1.8533×10^{-4}	2.5037×10^{-4}	2.0351×10^{-4}
1.8533×10^{-4}	1.0830×10^{-2}	1.0837×10^{-2}
1.0835×10^{-2}		

16 Node Scheme	17 Node Scheme	32 Node Scheme
3.4289×10^{-6}	3.4348×10^{-6}	3.4148×10^{-6}
4.5063×10^{-5}	5.8361×10^{-5}	5.0492×10^{-5}
5.5290×10^{-5}	$(8.6670 \pm j2.4399) \times 10^{-5}$	$(7.9651 \pm j4.1765) \times 10^{-5}$
1.4812×10^{-4}	1.2158×10^{-4}	1.6737×10^{-4}
1.9001×10^{-4}	1.0863×10^{-2}	1.0923×10^{-2}
1.0878×10^{-2}		

33 Node Scheme	64 Node Scheme	128 Node Scheme
3.4214×10^{-6}	3.4164×10^{-6}	3.4153×10^{-6}
4.4482×10^{-5}	4.4780×10^{-5}	4.4971×10^{-5}
$(8.3540 \pm j3.3537) \times 10^{-5}$	$(7.1522 \pm j5.3815) \times 10^{-5}$	$(7.4217 \pm j4.9218) \times 10^{-5}$
1.9140×10^{-4}	1.8939×10^{-4}	1.7236×10^{-4}
1.0903×10^{-2}	1.0942×10^{-2}	1.0931×10^{-2}

The two and three node models exhibit 8 unstable eigenvalues each, whereas the schemes with four and five nodes respectively have 12 unstable eigenvalues. The number of unstable eigenvalues is 10 for the schemes with 8 nodes onwards, which indicates consistency in capturing of stability property. Among them,

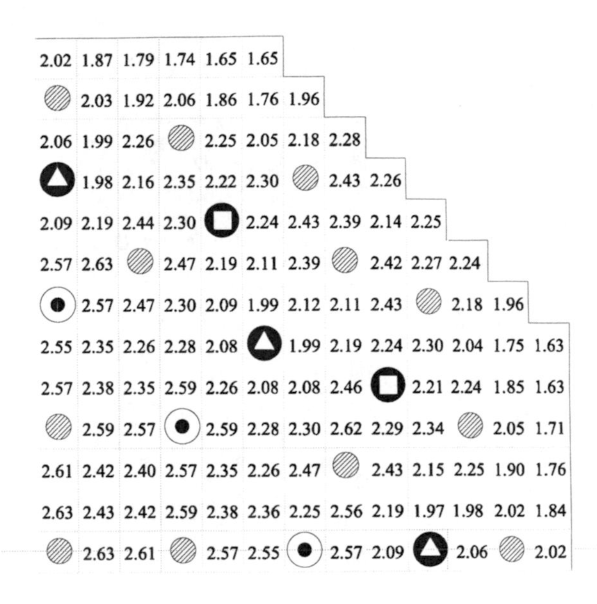

Fig. 2.1 Channel power distribution obtained from Finite Difference Model. Channel powers are expressed in MW (thermal) units.

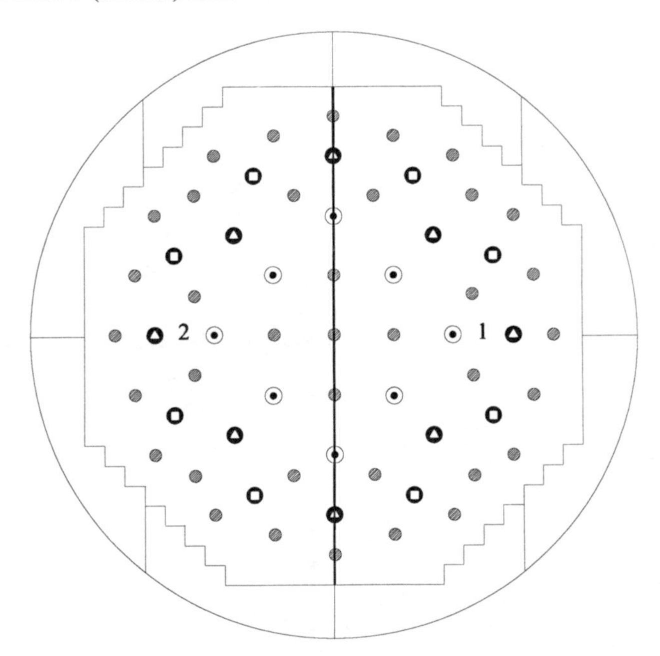

Fig. 2.2 Division of AHWR Core into Two Nodes.

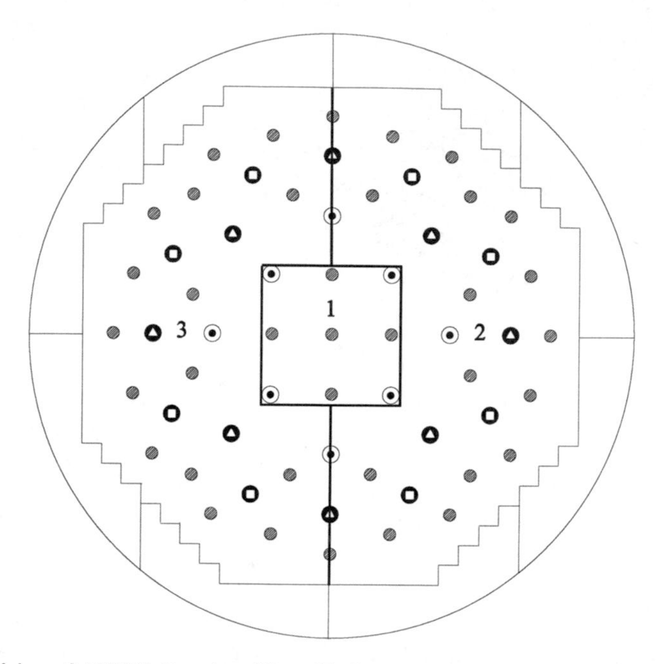

Fig. 2.3 Division of AHWR Core into Three Nodes.

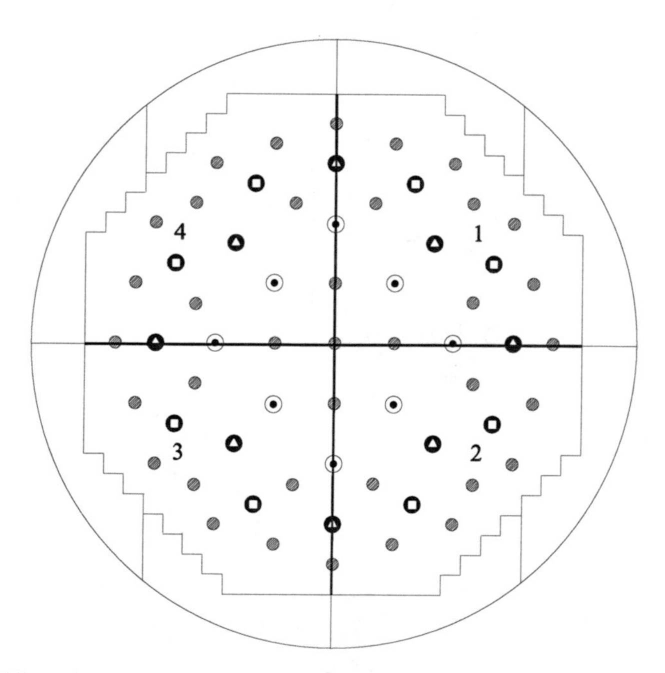

Fig. 2.4 Division of AHWR Core into Four Nodes.

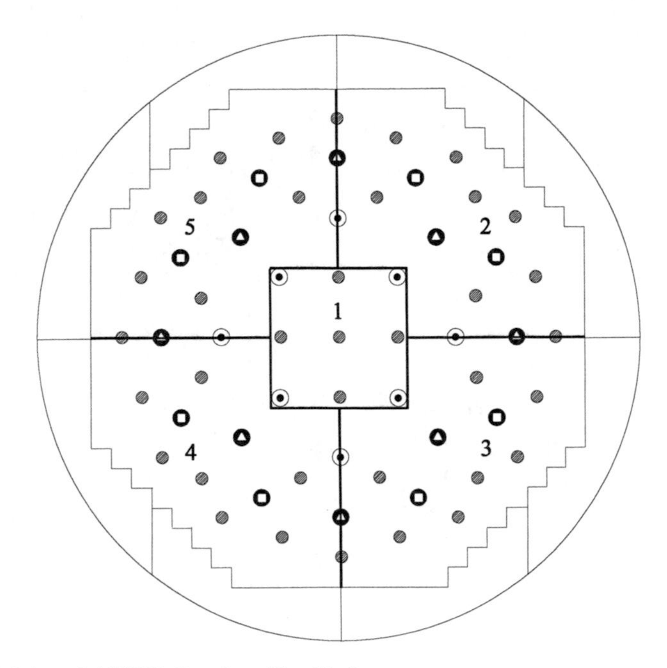

Fig. 2.5 Division of AHWR Core into Five Nodes.

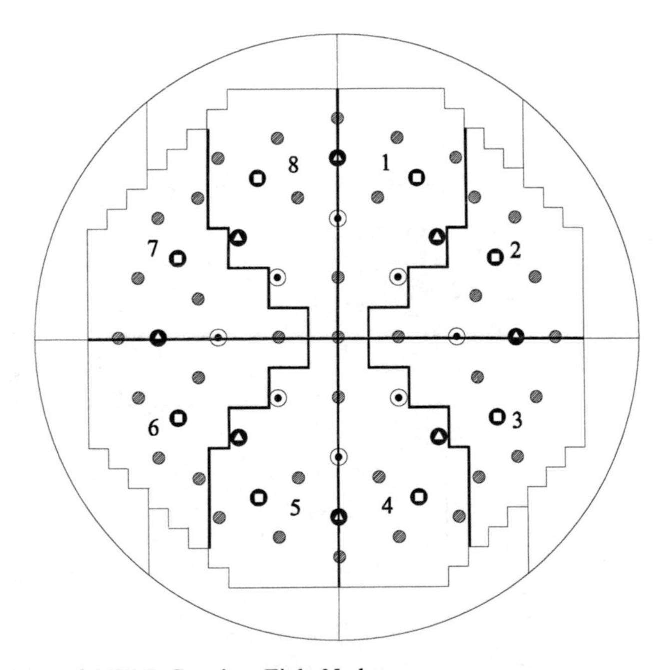

Fig. 2.6 Division of AHWR Core into Eight Nodes.

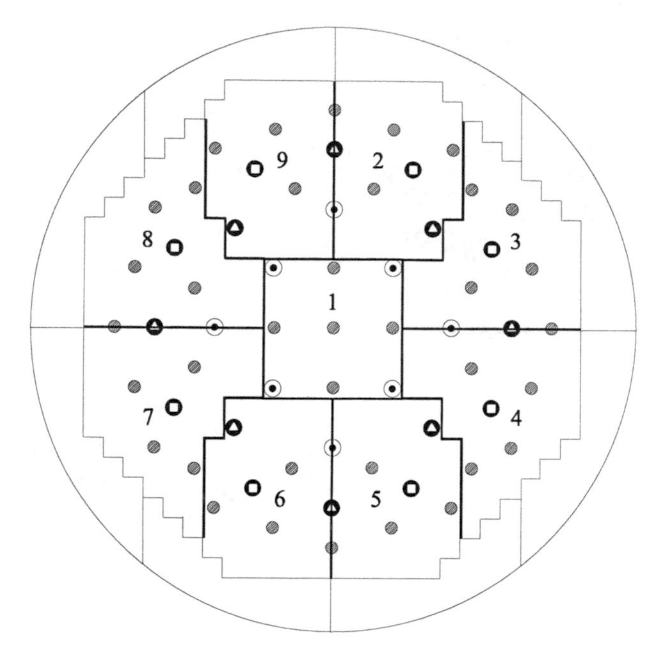

Fig. 2.7 Division of AHWR Core into Nine Nodes.

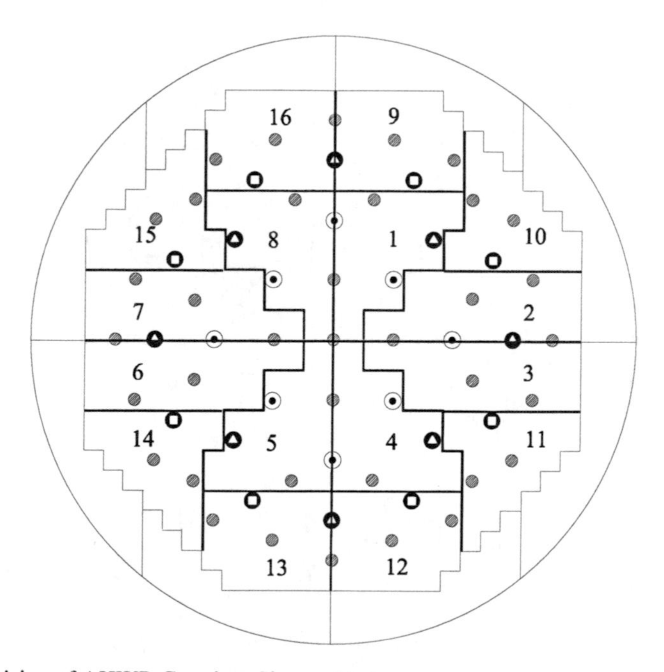

Fig. 2.8 Division of AHWR Core into Sixteen Nodes.

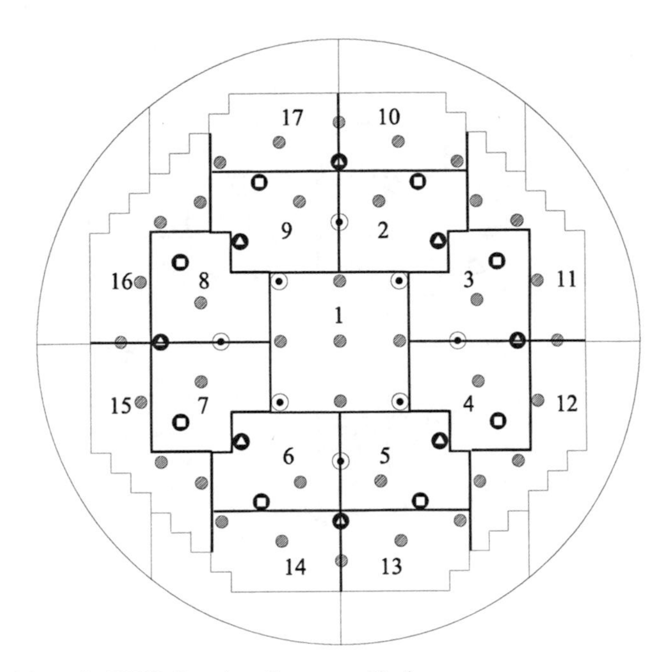

Fig. 2.9 Division of AHWR Core into Seventeen Nodes.

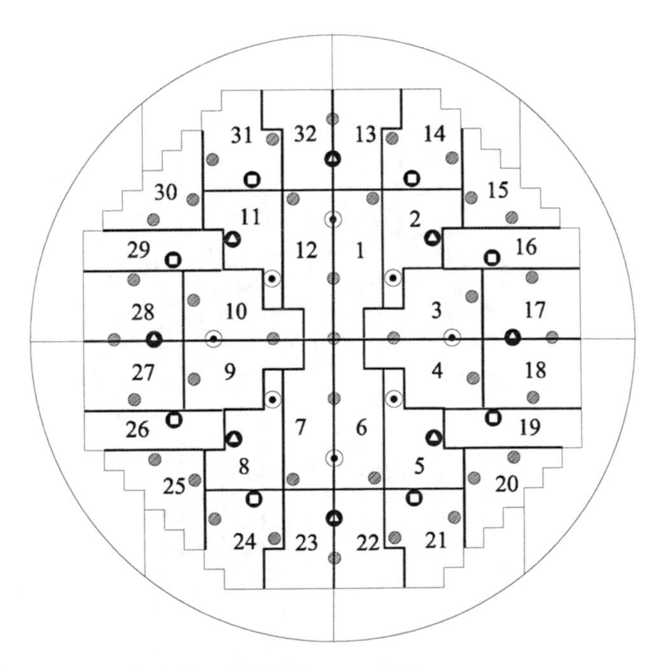

Fig. 2.10 Division of AHWR Core into Thirty two Nodes.

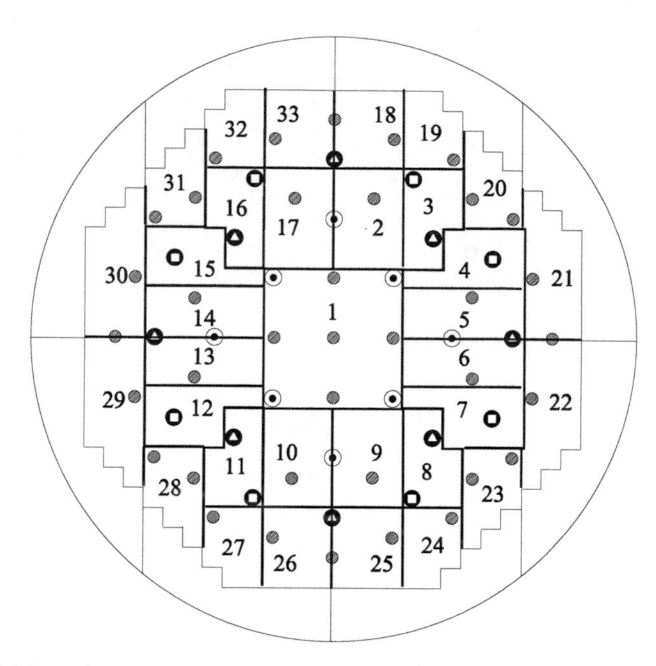

Fig. 2.11 Division of AHWR Core into Thirty three Nodes.

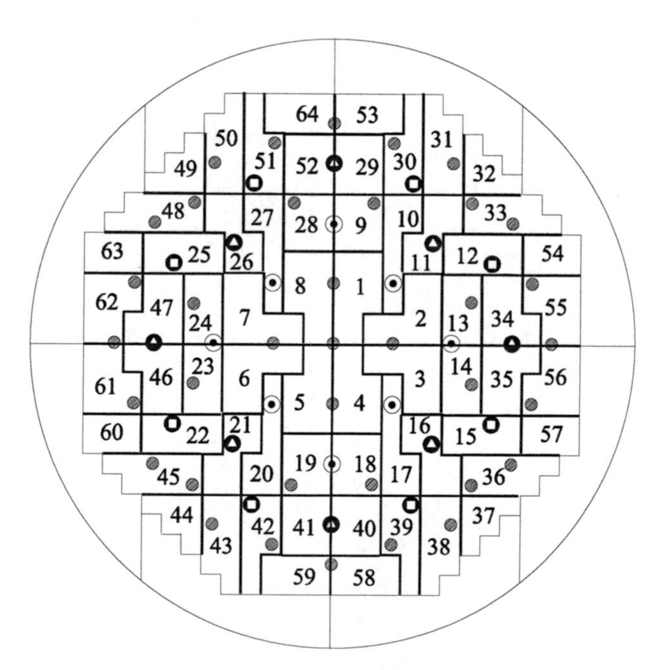

Fig. 2.12 Division of AHWR Core into Sixty four Nodes.

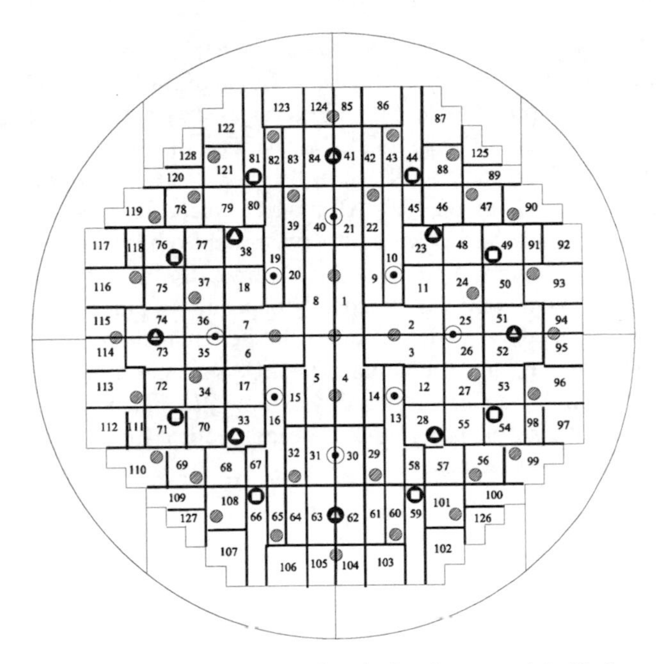

Fig. 2.13 Division of AHWR Core into One hundred and twenty eight Nodes.

however, schemes with 17 node onwards presents one complex conjugate pair among the unstable eigenvalues, which is missing in the 8 and 16 node schemes. In all the schemes, lower and upper bounds of the cluster of unstable eigenvalues are found to be more or less the same. Altogether it can be noticed that the number of unstable eigenvalues remain the same for the schemes with 17 nodes or more.

For each scheme, it was observed that the stable eigenvalues fall in three distinct clusters, with the real parts of first, second and third cluster ranging respectively between -10^{-6} to -10^{-4}, -10^{-2} to -10^{-1} and -10^{0} to -10^{2}. It was further noticed that the number of eigenvalues in the third cluster is equal to the number of nodes under consideration.

2.4.3 *Controllability Aspects*

For studying the controllability aspects, it was assumed that one RR in each quadrant of the reactor core is provided for automatic control under normal operating conditions. These RRs were chosen as those in lattice locations $E9$, $J21$, $R5$ and $V17$ (refer Fig. 1.1). System matrices A and B were formulated accordingly and controllability was assessed for each scheme. Controllability test of the (A,B) pair revealed that each of the models are controllable.

Controllability property was also assessed under all possible combinations of single as well as multiple failures of these RRs. Condition of unavailability of one/multiple RRs was imposed by forcing the corresponding column/columns of the input matrix B to all zeros. This simulates a condition in which the control signal to the respective RR/RRs is cut off. Each of the models exhibited full controllability under unavailability of any one RR. All models except for those corresponding to the four and five node schemes were fully controllable under simultaneous unavailability of all combinations of any two or even three RRs. However, under the event of simultaneous unavailability of diagonally opposite RRs (at $E9$ and $R5$, or at $J21$ and $V17$), the four and five node schemes exhibited 8 uncontrollable modes. Out of these, 4 modes were unstable. Four and five node schemes turned out to be uncontrollable under simultaneous unavailability of three RRs of any combination as well. It was noticed that controllability properties are identical for all schemes except for the four and five node schemes.

2.4.4 Observability Aspects

Initially, observability of each scheme was analyzed with only one output available, corresponding to that of the out–of–core ion chamber. It was established that each of the schemes are unobservable when only the total reactor power is being monitored. Subsequently, it was considered that the power in each quadrant of the core is also being monitored along with the total reactor power, in lines with the actual nuclear instrumentation system design. Under this assumption, each of the schemes turned out to be fully observable.

Furthermore, in lines with the controllability analysis, all possible combinations of single as well as multiple failures of the incore detectors (with and without the out–of–core detector) were also considered by forcing the respective row/rows of the output matrix M to all zeros. The results were identical to those of the controllability analysis. Barring for the 4 and 5 node schemes, all other schemes exhibited observability even when power in only any one quadrant is being monitored along with the total reactor power. Each model turned out to be unobservable in the event of unavailability of the total reactor power measurement, even when all four quadrant powers are being monitored. It was further noticed that observability properties are also identical for all schemes except for the four and five node schemes.

To summarize, controllability and observability properties remain uniform for schemes with more than 5 nodes. However, in order to achieve a more accurate channel power distribution, and in order to capture the stability features of a very high order model (including the complex conjugate pair of unstable eigenvalues), minimum 17 nodes are required.

In nutshell, among these different nodalization schemes considered, the one with 17 nodes as shown in Fig. 2.9 would be most suitable for AHWR, since it captures all the essential linear system properties. At the same time, this scheme also offers a satisfactory accuracy in steady state power distributions.

2.5 Thermal Hydraulics Model

In many applications of nuclear reactor analysis and control it may be necessary to consider the reactivity feedbacks due to various temperature coefficients and thermal hydraulics, for which it would be essential to consider dynamics of fuel and coolant temperatures, coolant void fraction *etc.* in the model. However, the extent of modeling of such phenomena is purely reactor and application specific. Let us consider the specific case of AHWR for developing a simplified thermal hydraulics model. Coolant temperature coefficient in AHWR is positive with a magnitude of $5.0 \times 10^{-5} \Delta k/k/^{o}C$. Its role is significant only during the reactor startup, since the coolant temperature remains constant at $285^{o}C$ during power operation. Fuel temperature reactivity feedback is negative, with a comparatively small coefficient of the order of $-28.70\mu k/^{o}C$. Moderator temperature coefficient is slightly positive, however, during normal reactor operation the moderator temperature remains almost constant thereby nullifying the significance of moderator temperature feedback. Due to their less significance, fuel, coolant and moderator temperature reactivity feedbacks are ignored. The gross average value of void reactivity coefficient, defined as change in reactivity with respect to the change in coolant voids, is negative for AHWR. This provides negative reactivity during overpower transients and loss of coolant accident conditions. In order to consider this reactivity feedback, it becomes essential to model the void reactivity, which requires two-phase coolant flow modeling and modeling of steam drums, *etc.* Following is a discussion on how to derive a simplified thermal hydraulics model for AHWR.

The Main Heat Transport (MHT) system of AHWR consists of 452 coolant channels in the reactor core, an equal number of tail pipes, 4 horizontal cylindrical steam drums, 16 downcomers, an inlet header and 452 inlet feeders. Coolant starts boiling in the reactor core by absorbing the fission heat. The steam-water mixture from the coolant channels flows to the steam drums through the tail pipes. Steam-water phase separation and feed water mixing takes place inside the steam drums, and the subcooled water flows back to the coolant channels through the downcomers, inlet header and inlet feeders. Fig. 2.14 shows a representative schematic of AHWR MHT. A simplified thermal hydraulics model of the MHT system of AHWR has been developed by evolving separate models for reactor core thermal hydraulics and for the steam drums, and afterwards clubbing them together, as detailed below.

2.5.1 Core Thermal Hydraulics

A thermal hydraulics model of the reactor core can be obtained assuming an equivalent coolant channel for each node, ignoring the pressure drops in downcomers, feeders and tail pipes, and taking uniform distribution of nodal power along the flow axis. Also the steam quality is considered to be uniformly increasing along the axial length in the channels after the point of onset of boiling.

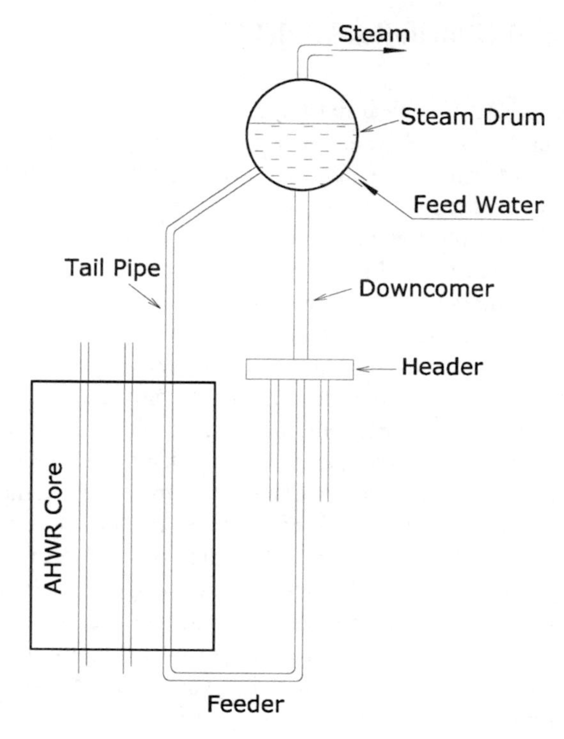

Fig. 2.14 Schematic of MHT System of AHWR.

The fraction of nodal power used for converting sub cooled water into saturated water is $Q_{1_h} = q_{d_h}(h_w - h_d)$ and the fraction of power available for boiling (converting the saturated water into steam) is $Q_{2_h} = Q_h - q_{d_h}(h_w - h_d)$. Hence, the boiling volume in the lumped channel of a node is:

$$V_{b_h} = V_{C_h} \frac{Q_h - q_{d_h}(h_w - h_d)}{Q_h}. \tag{2.37}$$

Application of the mass balance in the boiling section yields:

$$\frac{d}{dt}\left(\rho_s \overline{\alpha}_h V_{b_h} + \rho_w (1 - \overline{\alpha}_h) V_{b_h}\right) = q_{d_h} - q_{r_h} \tag{2.38}$$

where $\overline{\alpha}_h$ is the average void fraction in the node h, which is related to exit quality x_h as:

$$\overline{\alpha}_h = \frac{\rho_w}{\rho_w - \rho_s}\left[1 - \frac{\rho_s}{(\rho_w - \rho_s)x_h}\log\left(1 + \frac{\rho_w - \rho_s}{\rho_s}x_h\right)\right]. \tag{2.39}$$

Similarly, the application of the energy balance in the boiling section gives:

$$\frac{d}{dt}\left(\rho_s h_s \overline{\alpha}_h V_{b_h} + \rho_w h_w (1-\overline{\alpha}_h) V_{b_h} - P V_{b_h}\right) = Q_{2_h} + q_{d_h} h_w - (x_h h_c + h_w) q_{r_h}. \quad (2.40)$$

Multiplying (2.38) by $-(x_h h_c + h_w)$ and adding to (2.40), we get:

$$e_{vp_h}\frac{dP}{dt} + e_{vx_h}\frac{dx_h}{dt} = Q_h - q_{d_h}(h_w - h_d) - x_h h_c q_{d_h} \quad (2.41)$$

where e_{vp_h} and e_{vx_h} are given by:

$$e_{vp_h} = (\rho_w \frac{dh_w}{dP} - x_h h_c \frac{d\rho_w}{dP})(1-\overline{\alpha}_h) V_{b_h} \quad (2.42)$$

$$+((1-x_h) h_c \frac{d\rho_s}{dP} + \rho_s \frac{dh_s}{dP}) \overline{\alpha}_h V_{b_h} \quad (2.43)$$

$$+(\rho_s + (\rho_w - \rho_s) x_h) h_c V_{b_h} \frac{d\overline{\alpha}_h}{dP} - V_{b_h} \quad (2.44)$$

$$e_{vx_h} = ((1-x_h)\rho_s + x_h \rho_w) h_c V_{b_h} \frac{d\overline{\alpha}_h}{dx_h}. \quad (2.45)$$

2.5.2 Steam Drums

A simple lumped model of the steam drums is developed assuming that carry over and carry under effects are insignificant; a mixture of saturated water and steam enters the steam drum and subcooled water leaves steam drum into the reactor core; and average values of density and enthalpy of the water in the steam drum have been considered.

Conservation of mass and energy in the steam drum yields respectively [7]:

$$\frac{d}{dt}\left(\frac{(\rho_w + \rho_d) V_w}{2}\right) + \frac{d}{dt}(\rho_s V_s) = q_r + q_f - q_d - q_s \quad (2.46)$$

$$\frac{d}{dt}\left(\frac{(\rho_w u_w + \rho_d u_d) V_w}{2}\right) + \frac{d}{dt}(\rho_s u_s V_s) + m_d C_p \frac{dt_m}{dt} = x q_r h_s + (1-x) q_r h_w - q_d h_d + q_f h_f - q_s h_s \quad (2.47)$$

where;

$$x = \frac{\sum_{h=1}^{Z} x_h q_{r_h}}{\sum_{h=1}^{Z} q_{r_h}}; \quad q_r = \sum_{h=1}^{Z} q_{r_h}; \quad q_d = \sum_{h=1}^{Z} q_{d_h};$$

and ρ_s, ρ_w, ρ_d, h_s, h_w, h_d are functions of steam pressure, as determined from the steam table [30, 31]. Substituting internal energy $u = h - \frac{P}{\rho_d}$ and $V_d = V_s + V_w$, in (2.46) and (2.47), we have:

$$e_{pv}\frac{dV_w}{dt}+e_{pp}\frac{dP}{dt} = -\sum_{h=1}^{Z}\left(q_{d_h}-q_{r_h}\right)+q_f-q_s \tag{2.48}$$

$$e_{xv}\frac{dV_w}{dt}+e_{xp}\frac{dP}{dt} = xq_rh_s+(1-x)q_rh_w-q_dh_d+q_fh_f-q_sh_s \tag{2.49}$$

where e_{pv}, e_{pp}, e_{xv} and e_{xp} are given by:

$$\begin{aligned}
e_{pv} &= \frac{\rho_w+\rho_d}{2}-\rho_s \\
e_{pp} &= V_s\frac{d\rho_s}{dP}+V_w\frac{d(\rho_w+\rho_d)}{2dP} \\
e_{xv} &= \frac{1}{2}(d_wh_w+\rho_dh_d\,)-\rho_sh_s \\
e_{xp} &= \frac{1}{2}V_w(h_w\frac{d\rho_w}{dP}+\rho_w\frac{dh_w}{dP}+h_d\frac{d\rho_d}{dP}+\rho_d\frac{dh_d}{dP}) \\
&\quad +V_s(h_s\frac{d\rho_s}{dP}+\rho_s\frac{dh_s}{dP})-V_d+m_dC_p\frac{dt_s}{dP}
\end{aligned}$$

From (2.38), it follows that:

$$q_{d_h}-q_{r_h} = \overline{e}_{pp_h}\frac{dP}{dt}+\overline{e}_{px_h}\frac{dx_h}{dt} \tag{2.50}$$

where

$$\begin{aligned}
\overline{e}_{pp_h} &= V_{b_h}\left((1-\overline{\alpha}_h)\frac{d\rho_w}{dP}+\overline{\alpha}_h\frac{d\rho_s}{dP}-(\rho_w-\rho_s)\frac{d\overline{\alpha}_h}{dP}\right) \\
\overline{e}_{px_h} &= -V_{b_h}\left((\rho_w-\rho_s)\frac{\overline{\alpha}_h}{dx_h}\right).
\end{aligned}$$

Substituting (2.50) in (2.48), equations governing the dynamic model of the steam drum can be rewritten as:

$$e_{pv}\frac{dV_w}{dt}+\overline{e}_{pp}\frac{dP}{dt}+\sum_{h=1}^{Z}\overline{e}_{px_h}\frac{dx_h}{dt} = q_f-q_s \tag{2.51}$$

$$\begin{aligned}
e_{xv}\frac{dV_w}{dt}+e_{xp}\frac{dP}{dt} &= xq_rh_s+(1-x)q_rh_w-q_dh_d \\
&\quad +q_fh_f-q_sh_s
\end{aligned} \tag{2.52}$$

where $\overline{e}_{pp}=e_{pp}+\sum_{h=1}^{Z}\overline{e}_{pp_h}$. The coolant inlet flow, q_{d_h}, can be modeled considering the momentum balance equation, however for the sake of simplicity, it is modeled as a polynomial function obtained from the available data from a detailed thermal hydraulic model of AHWR, which correlates the reactor power to the coolant flow rate. Thus the flow rate in any node is represented as a function of normalised nodal power as:

$$q_{d_h} = \left(k_1 (\frac{Q_h}{Q_{h0}})^3 + k_2 (\frac{Q_h}{Q_{h0}})^2 + k_3 (\frac{Q_h}{Q_{h0}}) + k4 \right) q_{d_{h0}} \tag{2.53}$$

where $k1 = 0.2156$; $k2 = -0.5998$; $k3 = 0.48538$ and $k4 = 0.8988$.

Likewise, solving for the energy balance within the water volume of the steam drum yields

$$\frac{d}{dt} \left[\frac{(\rho_w u_w + \rho_d u_d) V_w}{2} \right] = (1-x) q_r h_w + q_f h_f - q_d h_d, \tag{2.54}$$

which can be manipulated to obtain

$$e_{vh} \frac{dV_w}{dt} + e_{ph} \frac{dP}{dt} + e_{xh} \frac{dh_d}{dt} = (1-x) q_r h_w + q_f h_f - q_d h_d, \tag{2.55}$$

where

$$e_{vh} = \frac{\rho_w h_w}{2} - P + \frac{\rho_d h_d}{2}, \tag{2.56}$$

$$e_{ph} = \frac{\rho_w V_w}{2} \frac{dh_w}{dP} + \frac{h_w V_w}{2} \frac{d\rho_w}{dP} - V_w + \frac{h_d V_w}{2} \frac{d\rho_d}{dP}, \tag{2.57}$$

$$e_{xh} = \frac{\rho_d V_w}{2}. \tag{2.58}$$

(2.41), (2.48), (2.49) and (2.55) govern the thermal hydraulics model of AHWR.

In applications involving the design of suitable controllers for spatial power distribution of AHWR, the thermal hydraulics model can be further simplified. It can reasonably be assumed that the steam drum level and pressure are being strictly regulated at their respective setpoints such that their time derivative terms are zero, since a dedicated secondary side control system termed as the overall plant control system (OPCS) regulates the steam drum level and pressure at their respective setpoints in AHWR. With this the thermal hydraulics model of AHWR reduces to

$$e_{vx_h} \frac{dx_h}{dt} = Q_h - q_{d_h} (h_w - h_d) - x_h h_c q_{d_h}, \tag{2.59}$$

$$e_{xh} \frac{dh_d}{dt} = q_f \left(\hat{k}_2 h_f - \hat{k}_1 \right) - q_d \left(\hat{k}_2 h_d - \hat{k}_1 \right), \tag{2.60}$$

where $\hat{k}_2 = \frac{h_s}{h_c}$ and $\hat{k}_1 = h_w \hat{k}_2$. Values of saturated enthalpies and densities are computed from steam tables [30, 31]. Further, the reactivity feedback due to coolant void fraction is given by

$$\rho_{h\alpha} = -5 \times 10^{-3} \left(9.2832 x_h^5 - 27.7192 x_h^4 + 31.7643 x_h^3 - 17.7389 x_h^2 + 5.2308 x_h + 0.0792 \right). \tag{2.61}$$

It may be noticed that under normal operating conditions, e_{xh} in (2.60) is a constant as denoted by (2.58). e_{vx_h} being a function of x_h and V_b varies slightly with the operating power level. However, solution of (2.45), (2.37), (2.39), (2.53) and the steady state form of (2.41) for variation of Q_h demonstrates that the maximum deviation in e_{vx_h}, which occurs in node 1, is about 2.8% for a change in Q_1 from 100% to 5%, as depicted in Fig. 2.15. Hence, for simplicity, e_{vx_h} can also assumed to be constant with values corresponding to those at full power operation, as listed in Table 2.5.

Table 2.5 Constant coefficients of thermal hydraulics model.

Node No.	e_{vx}
1	2.4406
2, 5, 6, 9	1.0909
3, 4, 7, 8	1.2160
10, 13, 14, 17	1.9861
11, 12, 15, 16	1.1677
$e_{xh} = 0.5114$	

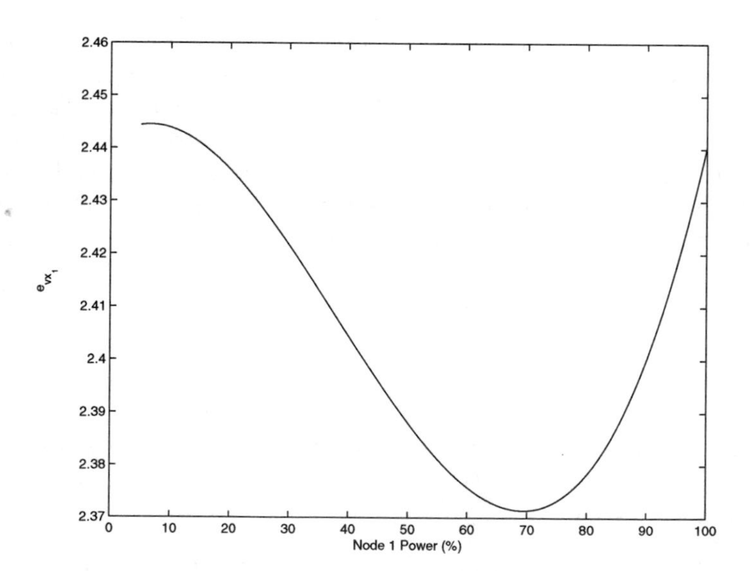

Fig. 2.15 Variations in e_{vx} with power level, in Node 1.

2.5.3 Steady State Response

Let $\overline{q}_{d_j}$, $j=1,\cdots,513$, denote the steady state channel flow distribution as shown in Fig. 2.16 under full power operating conditions, obtained from the detailed thermal hydraulics model. Also let $\overline{x}_j$, $j=1,\cdots,513$, denote the corresponding channel exit quality, as shown in Fig. 2.17. Taking these as the reference, steady state values of reference exit quality of each of the 17 nodes of the reactor core are obtained as

$$\hat{x}_h = \frac{\sum_{j\in h}\overline{q}_{d_j}\overline{x}_j}{\sum_{j\in h}\overline{q}_{d_j}}, \quad h=1\cdots 17. \tag{2.62}$$

Table 2.6 compares these reference exit qualities with x_{h0}, the steady state exit qualities under full power operating conditions computed from the steady state form of (2.59), which shows that the maximum error is within 3.5%.

Table 2.6 Comparison of Reference and Computed Exit Quality for the 17 Nodes under Full Power Operation.

Node No., h	$\hat{x}_h$	x_{h0}	% Error
1	0.2340	0.2385	−1.92
2, 5, 6, 9	0.1979	0.1931	2.43
3, 4, 7, 8	0.2012	0.2079	−3.33
10, 13, 14, 17	0.1992	0.2047	−2.76
11, 12, 15, 16	0.1898	0.1962	−3.37

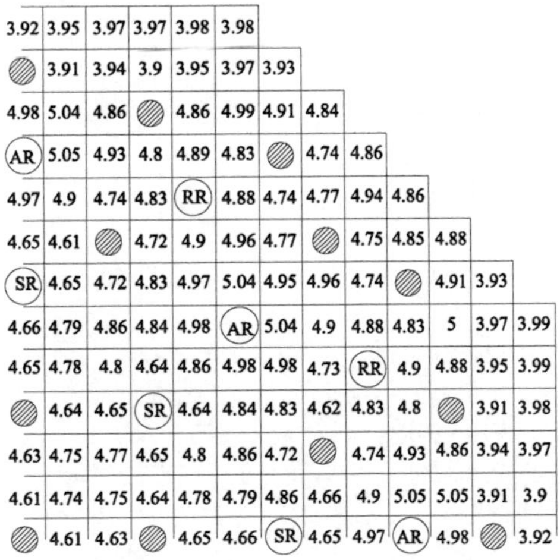

3.92	3.95	3.97	3.97	3.98	3.98							
	3.91	3.94	3.9	3.95	3.97	3.93						
4.98	5.04	4.86		4.86	4.99	4.91	4.84					
AR	5.05	4.93	4.8	4.89	4.83		4.74	4.86				
4.97	4.9	4.74	4.83	RR	4.88	4.74	4.77	4.94	4.86			
4.65	4.61		4.72	4.9	4.96	4.77		4.75	4.85	4.88		
SR	4.65	4.72	4.83	4.97	5.04	4.95	4.96	4.74		4.91	3.93	
4.66	4.79	4.86	4.84	4.98	AR	5.04	4.9	4.88	4.83	5	3.97	3.99
4.65	4.78	4.8	4.64	4.86	4.98	4.98	4.73	RR	4.9	4.88	3.95	3.99
	4.64	4.65	SR	4.64	4.84	4.83	4.62	4.83	4.8		3.91	3.98
4.63	4.75	4.77	4.65	4.8	4.86	4.72		4.74	4.93	4.86	3.94	3.97
4.61	4.74	4.75	4.64	4.78	4.79	4.86	4.66	4.9	5.05	5.05	3.91	3.9
	4.61	4.63		4.65	4.66	SR	4.65	4.97	AR	4.98		3.92

Fig. 2.16 Channel Flow Distribution under Full Power, obtained from Detailed Thermal Hydraulics Model.

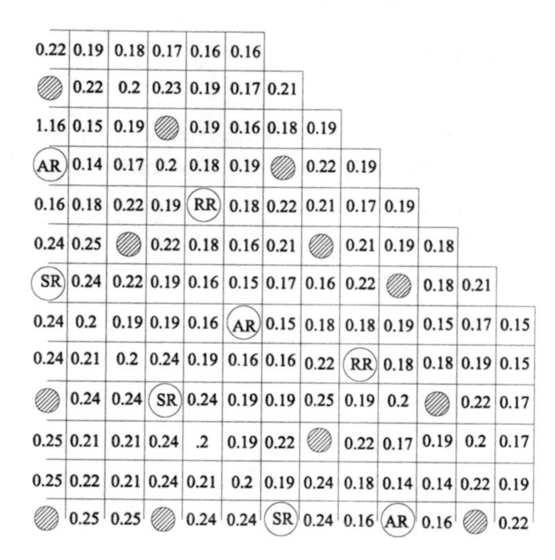

Fig. 2.17 Channel Quality Distribution under Full Power, obtained from Detailed Thermal Hydraulics Model.

2.6 Coupled Neutronics–Thermal Hydraulics Model

Having obtained the neutronics and thermal hydraulics models separately, now it is needed to couple the two to obtain a complete model of the reactor [100]. Clubbing the neutronics and thermal hydraulics equations together, the coupled neutronics–thermal hydraulics model of the AHWR is obtained. It consists of the following set of equations:

$$\frac{dQ_h}{dt} = -\alpha_{hh}\frac{Q_h}{\ell_h} + \sum_{k=1}^{17} \alpha_{kh}\frac{Q_k}{\ell_h} + (\rho_h - \beta)\frac{Q_h}{\ell_h} + \sum_{i=1}^{m_d} \lambda_i C_{ih}, \tag{2.63}$$

$$\frac{dC_{ih}}{dt} = \frac{\beta_i}{\ell_h} Q_h - \lambda_i C_{ih}, \quad i = 1, 2, \cdots m_d, \tag{2.64}$$

$$\frac{dI_h}{dt} = \gamma_I \Sigma_{fh} Q_h - \lambda_I I_h, \tag{2.65}$$

$$\frac{dX_h}{dt} = \gamma_x \Sigma_{fh} Q_h + \lambda_I I_h - (\lambda_x + \overline{\sigma}_{xh} Q_h) X_h \tag{2.66}$$

$$e_{vx_h}\frac{dx_h}{dt} = Q_h - q_{d_h}(h_w - h_d) - x_h h_c q_{d_h}, \tag{2.67}$$

$$e_{xh}\frac{dh_d}{dt} = q_f\left(\hat{k}_2 h_f - \hat{k}_1\right) - q_d\left(\hat{k}_2 h_d - \hat{k}_1\right), \tag{2.68}$$

$$\frac{dH_j}{dt} = K v_j, \quad h = 1, 2, \cdots 17, \quad j = 2, 4, 6, 8, \tag{2.69}$$

where q_{d_h} is a function of Q_h as given by (2.53), and the neutronic parameters and the coupling coefficients as listed in Tables 2.7 and 2.8 respectively.

Table 2.7 Nodal Volumes and Cross–Sections.

Node No.	Volume (m^3)	Σ_f (cm^{-1})	Σ_a (cm^{-1})
1	8.6822	2.6657×10^{-3}	6.9514×10^{-3}
2, 5, 6, 9	5.4042	2.3898×10^{-3}	6.6828×10^{-3}
3, 4, 7, 8	5.1384	2.5325×10^{-3}	6.7898×10^{-3}
10, 13, 14, 17	4.4297	2.5665×10^{-3}	6.8991×10^{-3}
11, 12, 15, 16	5.5814	2.5665×10^{-3}	6.8991×10^{-3}

Table 2.8 Coupling Coefficients for the AHWR Model.

$$\alpha_{1,1} = 3.1567 \times 10^{-2}$$
$$\alpha_{2,2} = \alpha_{5,5} = \alpha_{6,6} = \alpha_{9,9} = 5.4918 \times 10^{-2}$$
$$\alpha_{3,3} = \alpha_{4,4} = \alpha_{7,7} = \alpha_{8,8} = 6.2052 \times 10^{-2}$$
$$\alpha_{10,10} = \alpha_{13,13} = \alpha_{14,14} = \alpha_{17,17} = 3.8351 \times 10^{-2}$$
$$\alpha_{11,11} = \alpha_{12,12} = \alpha_{15,15} = \alpha_{16,16} = 4.3567 \times 10^{-2}$$
$$\alpha_{1,2} = \alpha_{1,5} = \alpha_{1,6} = \alpha_{1,9} = 6.5746 \times 10^{-3}$$
$$\alpha_{1,3} = \alpha_{1,4} = \alpha_{1,7} = \alpha_{1,8} = 6.5204 \times 10^{-3}$$
$$\alpha_{2,1} = \alpha_{5,1} = \alpha_{6,1} = \alpha_{9,1} = 4.5833 \times 10^{-3}$$
$$\alpha_{3,1} = \alpha_{4,1} = \alpha_{7,1} = \alpha_{8,1} = 4.3309 \times 10^{-3}$$
$$\alpha_{2,3} = \alpha_{5,4} = \alpha_{6,7} = \alpha_{9,8} = 1.3044 \times 10^{-2}$$
$$\alpha_{3,2} = \alpha_{4,5} = \alpha_{7,6} = \alpha_{8,9} = 1.2428 \times 10^{-2}$$
$$\alpha_{3,4} = \alpha_{4,3} = \alpha_{7,8} = \alpha_{8,7} = 1.6097 \times 10^{-2}$$
$$\alpha_{2,9} = \alpha_{5,6} = \alpha_{6,5} = \alpha_{9,2} = 1.0445 \times 10^{-2}$$
$$\alpha_{2,10} = \alpha_{5,13} = \alpha_{6,14} = \alpha_{9,17} = 2.3481 \times 10^{-2}$$
$$\alpha_{3,11} = \alpha_{4,12} = \alpha_{7,15} = \alpha_{8,16} = 2.7555 \times 10^{-2}$$
$$\alpha_{10,2} = \alpha_{13,5} = \alpha_{14,6} = \alpha_{17,9} = 1.9198 \times 10^{-2}$$
$$\alpha_{11,3} = \alpha_{12,4} = \alpha_{15,7} = \alpha_{16,8} = 2.9941 \times 10^{-2}$$
$$\alpha_{10,17} = \alpha_{17,10} = \alpha_{11,12} = \alpha_{12,11} = \alpha_{13,14} = \alpha_{14,13} = \alpha_{15,160} = \alpha_{16,15} = 9.9912 \times 10^{-3}$$

Rest all $\alpha_{i,j} = 0$.

In order to account for reactivity variations due to internal feedbacks and control devices, the reactivity term ρ_h in (2.63) is expressed as

$$\rho_h = \rho_{hX} + \rho_{h_u} + \rho_{h\alpha}, \tag{2.70}$$

where ρ_{h_u} is the reactivity introduced by the control rods, ρ_{hX} is the reactivity feedback due to xenon and ρ_{h_f} is the reactivity feedback due to coolant void fraction. The reactivity contributed by the movement of the RRs around their equilibrium positions is expressed as

$$\rho_{h_u} = \begin{cases} (-10.234 H_h + 676.203) \times 10^{-6} & \text{if } h = 2,4,6 \text{ or } 8 \\ 0 \text{ elsewhere.} \end{cases} \tag{2.71}$$

The xenon reactivity feedback in node h can be expressed as

$$\rho_{hX} = -\frac{\overline{\sigma}_{xh} X_h}{\Sigma_{ah}}. \tag{2.72}$$

The reactivity feedback due to void fraction, ρ_{h_f}, is given by

$$\rho_{h\alpha} = -5 \times 10^{-3} \Big(9.2832x_h^5 - 27.7192x_h^4 + 31.7643x_h^3 - 17.7389x_h^2 + 5.2308x_h + 0.0792\Big) . \tag{2.73}$$

2.6.1 Steady State Operation

Under steady state full power operation, the nodal power levels and coolant flow rates are constants as given in Table 2.9. The equilibrium position of the all the RRs is 66.1 % inside the core [90]. The coolant enters the core at a temperature of $260^o C$ and the feed water enters the steam drum at $130^o C$. The operating pressure of the main heat transport system of AHWR is 7 MPa. Steady state feed water enthalpy h_d, incoming coolant enthalpy h_d and saturated enthalpies of water and steam, corresponding to this pressure, are obtained from the steam table. Equilibrium values of other variables like delayed neutron precursors, iodine and xenon concentrations, core exit qualities, feed flow rate *etc.* can readily be computed from the steady state forms of the respective equations (2.63)–(2.68).

Table 2.9 Distribution of Nodal Powers and Coolant Flow rates under Full Power Operation.

Node No.	Steady State Values of	
	Power (MW Thermal)	Coolant Flow rate (kg/s)
1	91.8743	187.32
2, 5, 6, 9	54.9991	130.20
3, 4, 7, 8	55.7410	125.38
10, 13, 14, 17	42.6967	97.06
11, 12, 15, 16	53.7146	125.78
Total	920.480	2101.0

2.6.2 Linearization of Model Equations

The set of equations (2.63)–(2.69) can be linearized around the steady state conditions $(Q_{h0}, C_{h0}, I_{h0}, X_{h0}, H_{j0}, x_{h0}, h_{d0})$ just described. Define the state vector as

$$z = \left[z_H^T \; z_X^T \; z_I^T \; \delta h_d \; z_C^T \; z_x^T \; z_Q^T \right]^T , \tag{2.74}$$

where

$$z_H = \begin{bmatrix} \delta H_2 \; \delta H_4 \; \delta H_6 \; \delta H_8 \end{bmatrix}^T, \; z_X = \begin{bmatrix} \frac{\delta X_1}{X_{1_0}} \cdots \frac{\delta X_{17}}{X_{17_0}} \end{bmatrix}^T,$$
$$z_I = \begin{bmatrix} \frac{\delta I_1}{I_{1_0}} \cdots \frac{\delta I_{17}}{I_{17_0}} \end{bmatrix}^T, z_C = \begin{bmatrix} \frac{\delta C_1}{C_{1_0}} \cdots \frac{\delta C_{17}}{C_{17_0}} \end{bmatrix}^T,$$
$$z_x = \begin{bmatrix} \frac{\delta x_1}{x_{1_0}} \cdots \frac{\delta x_{17}}{x_{17_0}} \end{bmatrix}^T, z_Q = \begin{bmatrix} \frac{\delta Q_1}{Q_{1_0}} \cdots \frac{\delta Q_{17}}{Q_{17_0}} \end{bmatrix}^T,$$

in which δ denotes the deviation from respective steady state value of the variable. Likewise define the input vectors as

$$u_1 = \begin{bmatrix} \delta v_2 \; \delta v_4 \; \delta v_6 \; \delta v_8 \end{bmatrix}^T, \; u_2 = \delta q_f,$$

where u_1 consisting of the control signals to the regulating rod drives is the input being used for reactor power control, and u_2 which denotes the changes in the feed-water flow rate is treated as a disturbance.

Also define the output vector as

$$y = \begin{bmatrix} y_g \; y_1 \; \ldots \; y_Z \end{bmatrix}^T,$$

where $y_g = \sum_{i=1}^{Z} \frac{\delta Q_i}{\sum_{j=1}^{Z} Q_{j0}}$ and $y_i = \frac{\delta Q_i}{Q_{i0}}$ correspond to the normalized total reactor power and the nodal powers respectively. Then the system given by (2.63)–(2.69) can be expressed in standard linear state space form as

$$\begin{aligned} \dot{z} &= Az + B_1 u_1 + B_2 u_2 \\ y &= Mz. \end{aligned} \tag{2.75}$$

The characteristic matrix A is of order $Z(m_d + 4) + 5$, expressed as

$$A = \begin{bmatrix} 0 & 0 & 0 & 0 & 0 & 0 & 0 \\ 0 & A_{XX} & A_{XI} & 0 & 0 & 0 & A_{XQ} \\ 0 & 0 & A_{II} & 0 & 0 & 0 & A_{IQ} \\ 0 & 0 & 0 & A_{hh} & 0 & 0 & A_{hQ} \\ 0 & 0 & 0 & 0 & A_{CC} & 0 & A_{CQ} \\ 0 & 0 & 0 & A_{xh} & 0 & A_{xx} & A_{xQ} \\ A_{QH} & A_{QX} & 0 & 0 & A_{QC} & A_{Qx} & A_{QQ} \end{bmatrix} \tag{2.76}$$

where the first row corresponds to 4 rows of zeros, and 0 stands for null matrix of appropriate dimensions. Remaining sub blocks are given by:

$$A_{QQ}(i,j) = \begin{cases} \frac{1}{\ell}\left(-\sum_{k=1}^{Z} \alpha_{ki} \frac{Q_{k0}}{Q_{i0}} - \beta \right) & \text{if } i = j \\ \frac{1}{\ell} \alpha_{ji} \cdot \frac{Q_{j0}}{Q_{i0}} & \text{if } i \neq j \end{cases}$$

$$A_{QC} = \frac{\beta}{\ell} E_Z,$$

$$A_{QX} = -\frac{1}{\ell}diag.\left[\frac{\overline{\sigma}_{X1}}{\Sigma_{a1}}X_{1_0} \ \frac{\overline{\sigma}_{X2}}{\Sigma_{a2}}X_{2_0}\cdots\frac{\overline{\sigma}_{xZ}}{\Sigma_{XZ}}X_{Z_0}\right],$$

$$A_{Qx} = \frac{1}{\ell}diag.\left[k_{\alpha_1}Q_{1_0} \ k_{\alpha_2}Q_{2_0}\cdots k_{\alpha_Z}Q_{Z_0}\right];$$

where $k_{\alpha_i} = -5\times10^{-3}\left(46.4162x_{i0}^4 - 110.8787x_{i0}^3 + 95.229x_{i0}^2 - 35.4779x_{i0} + 5.2308\right)$,

$$A_{CQ} = \lambda E_Z,$$

$$A_{CC} = -A_{CQ},$$

$$A_{IQ} = \lambda_I E_Z \text{ and } A_{II} = -A_{IQ},$$

$$A_{XQ} = diag.\left[\lambda_X - \lambda_I\frac{I_{1_0}}{X_{1_0}} \ \lambda_X - \lambda_I\frac{I_{2_0}}{X_{2_0}}\cdots\lambda_X - \lambda_I\frac{I_{Z_0}}{X_{Z_0}}\right],$$

$$A_{XX} = -diag.\left[\lambda_X + \overline{\sigma}_{X1}Q_{1_0} \ \lambda_X + \overline{\sigma}_{X2}Q_{2_0}\cdots\lambda_X + \overline{\sigma}_{XZ}Q_{Z_0}\right],$$

$$A_{XI} = \lambda_I diag.\left[\frac{I_{1_0}}{X_{1_0}} \ \frac{I_{2_0}}{X_{2_0}}\cdots\frac{I_{Z_0}}{X_{Z_0}}\right],$$

$$A_{hh} = -\hat{k_1}\frac{q_{d0}}{e_{xh}h_w},$$

$$A_{hQ} = \frac{\hat{k}_1}{e_{xh}}(3k_1 + 2k_2 + k_3)\left(\frac{1}{h_{d0}} - \frac{1}{h_w}\right)\left[q_{d_{10}} \ q_{d_{20}} \ \cdots q_{d_{Z0}}\right],$$

$$A_{xh} = \left[\frac{q_{d_{10}}h_{d0}}{e_{vx_1}x_{10}} \ \frac{q_{d_{20}}h_{d0}}{e_{vx_2}x_{20}} \ \cdots \ \frac{q_{d_{Z0}}h_{d0}}{e_{vx_Z}x_{Z0}}\right]^T,$$

$$A_{xx} = -h_c diag.\left[\frac{q_{d_{10}}}{e_{vx_1}} \ \frac{q_{d_{20}}}{e_{vx_2}} \ \cdots \ \frac{q_{d_{Z0}}}{e_{vx_Z}}\right];$$

$$A_{xQ} = diag.\left[a_{xq_1} \ a_{xq_2} \ \ldots \ a_{xq_Z}\right];$$

where $a_{xq_i} = \dfrac{1}{e_{vx_i}x_{i0}}\left(Q_{i_0} - (h_w - h_{d0} + x_{i0}h_c)\,q_{d_{i0}}\,(3k_1 + 2k_2 + k_3)\right)$,

$$A_{QH}(i,j) = \begin{cases} \frac{-10.23\times10^{-6}Q_{i_0}}{\ell} & \text{for } i = 2,4,6,8; \ j = i/2\,. \\ 0 \text{ elsewhere} \end{cases}$$

Matrix B_1, of dimension $(Z(m_d+4)+5)\times 4$, is as follows:

$$B_1 = \left[B_H^T \ 0\,0\,0\,0\,0\,0\right]^T; \tag{2.77}$$

where B_H is a diagonal matrix of dimension 4×4, with K as diagonal entries.

Matrix B_2 is of dimension $(Z(m_d+4)+5)\times 1$ with $b_2 = \hat{k}_1 q_{f0}\left(\frac{h_{f0}-1}{e_{xh}h_{d0}}\right)$ on the $(2Z+5)^{th}$ row and remaining all entries zero.

Matrix M is of dimension $(Z+1)\times(Z(m_d+4)+5)$, given by

$$M = \left[M_1 \ \ M_2\right], \tag{2.78}$$

where M_1 is a null matrix and

$$M_2 = \begin{bmatrix} \frac{Q_{1_0}}{\sum_{j=1}^{Z} Q_{j0}} & \frac{Q_{2_0}}{\sum_{j=1}^{Z} Q_{j0}} & \cdots & \frac{Q_{Z_0}}{\sum_{j=1}^{Z} Q_{j0}} \\ 1 & 0 & \ldots & 0 \\ 0 & 1 & \ldots & 0 \\ \vdots & \vdots & \ddots & \vdots \\ 0 & 0 & \ldots & 1 \end{bmatrix}.$$

2.6.3 *Effect of Thermal Hydraulics on System Characteristics*

It may be recknoed that the nodalization scheme was devised by considering the neutronics model only. At this stage, when the thermal hydraulics model is coupled to the neutronics model, it is essential to investigate the impact of this coupling on the system characterestics and nodalization scheme. Table 2.10 lists the eigenvalues of the matrix A of the system given in (2.75) for each of the nodalization schemes discussed in Sec 2.4. On comparison with Table 2.4, it can be observed that the number of unstable eigenvalues remain unaffected with the thermal hydraulics feedback. However, the unstable eigenvalues tend to shift slightly towards the origin in complex s–plane, which implies that the void reactivity feedback tends to stabilize the system. It can also be verified that the location of unstable eigenvalues remain more or less similar for schemes with 17 nodes onwards. The properties like controllability and observability also remain unaffected with the void reactivity feedback. This gives confidence that the selected nodalization scheme is valid even with thermal hydraulics coupling.

2.7 Transient Response

Once the model is derived, it is very much essential to validate the model for the transient response characteristics. This needs to be done by simulating the dynamic response of the model for few typical transients and comparing the response with a benchmark response. To illustrate the dynamic behavior of the AHWR model, open loop response of the 17 node model under a few control relevant transients is presented here. Response obtained from the finite difference model is taken as the reference solution for validating the response of the nodal model. In all simulations, the reactor was assumed to be initially operating on full power. It was also assumed that initially each RR is equally at 66.1% in position, ARs are fully in and SRs and SORs are fully out, which is the critical configuration of the equilibrium core of AHWR.

Table 2.10 Unstable Eigenvalues Exhibited by Different Schemes, With Thermal Hydraulics Coupling (In each scheme, there are also four eigenvalues at origin not listed here).

2 Node Scheme	3 Node Scheme	4 Node Scheme
3.9281×10^{-6}	3.8981×10^{-6}	3.8881×10^{-6}
2.0207×10^{-5}	2.9123×10^{-5}	$(2.8512 \pm j1.0371) \times 10^{-5}$
3.9019×10^{-5}	3.4225×10^{-5}	$(1.3915 \pm j3.4129) \times 10^{-5}$
8.1019×10^{-3}	8.0219×10^{-3}	3.9288×10^{-5}
		2.0219×10^{-5}
		7.9194×10^{-3}
5 Node Scheme	**8 Node Scheme**	**9 Node Scheme**
3.8715×10^{-6}	3.9276×10^{-6}	3.8271×10^{-6}
$(3.2934 \pm j1.9224) \times 10^{-5}$	$(6.0201 \pm j1.2712) \times 10^{-5}$	$(5.9237 \pm j1.7012) \times 10^{-5}$
$(2.0170 \pm j2.6271) \times 10^{-5}$	2.7212×10^{-5}	1.9218×10^{-5}
$(1.0218 \pm j1.9001) \times 10^{-5}$	2.9225×10^{-5}	2.0212×10^{-5}
8.9279×10^{-3}	8.9918×10^{-3}	8.9220×10^{-3}
16 Node Scheme	**17 Node Scheme**	**32 Node Scheme**
3.9014×10^{-6}	3.9654×10^{-6}	3.9124×10^{-6}
$(6.2412 \pm j2.0239) \times 10^{-5}$	$(8.8268 \pm j1.8656) \times 10^{-5}$	$(8.7915 \pm j1.4397) \times 10^{-5}$
1.2314×10^{-5}	$(8.0470 \pm j2.4129) \times 10^{-5}$	$(7.9012 \pm j2.1578) \times 10^{-5}$
1.9917×10^{-5}	7.4551×10^{-3}	7.3381×10^{-3}
8.0216×10^{-3}		
33 Node Scheme	**64 Node Scheme**	**128 Node Scheme**
3.9248×10^{-6}	3.9014×10^{-6}	3.9302×10^{-6}
$(8.8124 \pm j1.8273) \times 10^{-5}$	$(8.8006 \pm j1.7951) \times 10^{-5}$	$(8.7992 \pm j1.8012) \times 10^{-5}$
$(8.1298 \pm j2.9517) \times 10^{-5}$	$(7.8541 \pm j1.5841) \times 10^{-5}$	$(8.0215 \pm j2.1358) \times 10^{-5}$
7.3991×10^{-3}	7.4158×10^{-3}	7.4215×10^{-3}

2.7.1 Simultaneous Movement of RRs

Simultaneous movement of 4 RRs, located in nodes 2, 4, 6 and 8 respectively, was considered to simulate a total power transient. A control signal of 1V was applied to the respective RR drives, under which these RRs moved linearly into the reactor core. When they were displaced in by 4.17% from the nominal positions, control signals were made zero in order to hold the RRs at their new positions. After a small interval of time, these RRs were driven out linearly under a control signal of -1V back to their nominal positions. After another short interval of time, an outward movement of these RRs followed by inward movement back to nominal position was simulated. Fig. 2.18 shows the position of these RRs during the transient. All other reactivity devices were kept stationary at their respective nominal positions.

Nodal power levels for this transient were determined by solving (2.21) – (2.25). Reference solution for the same transient was obtained from the detailed three–dimensional finite difference computations based on a quasi–static approach, with shape calculations performed at every 2.5 s besides at time $t = 0$ and amplitude

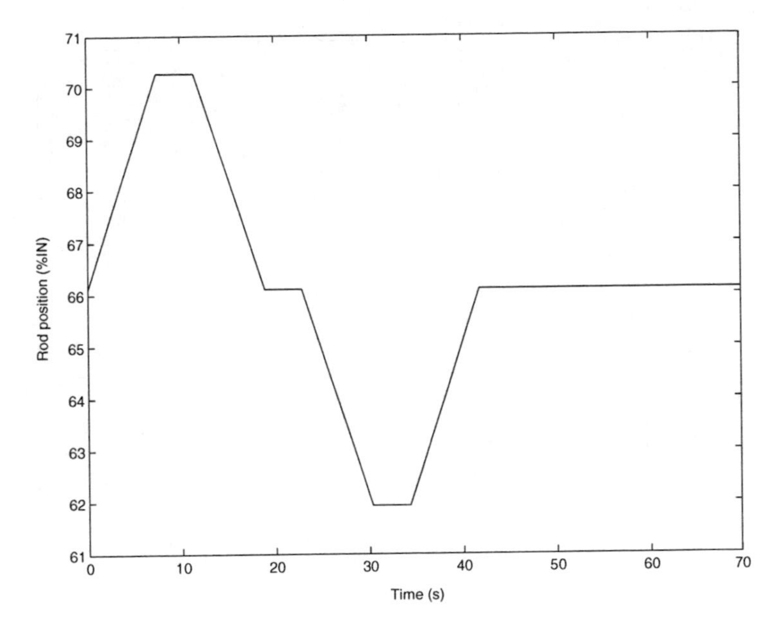

Fig. 2.18 Position of RRs during the Total Power Transient.

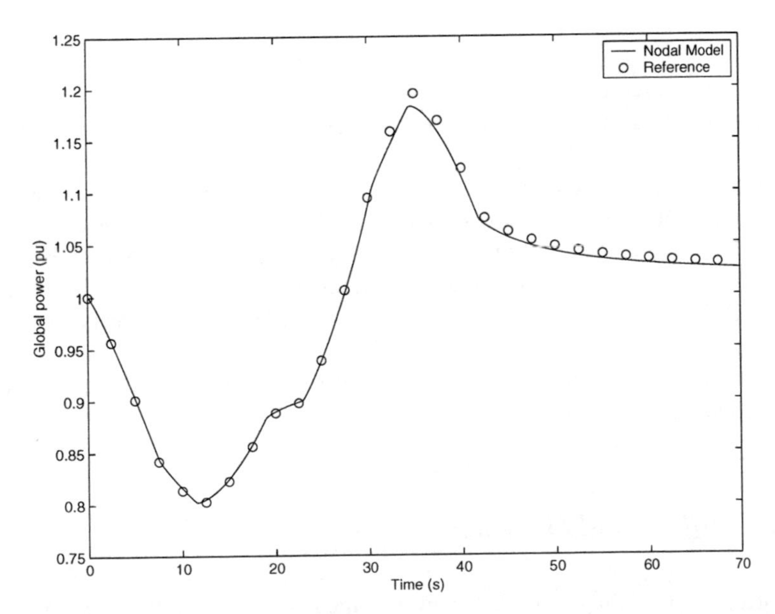

Fig. 2.19 Comparison of the Nodal Model Response with the Reference Solution during the Simultaneous Movement of 4 RRs.

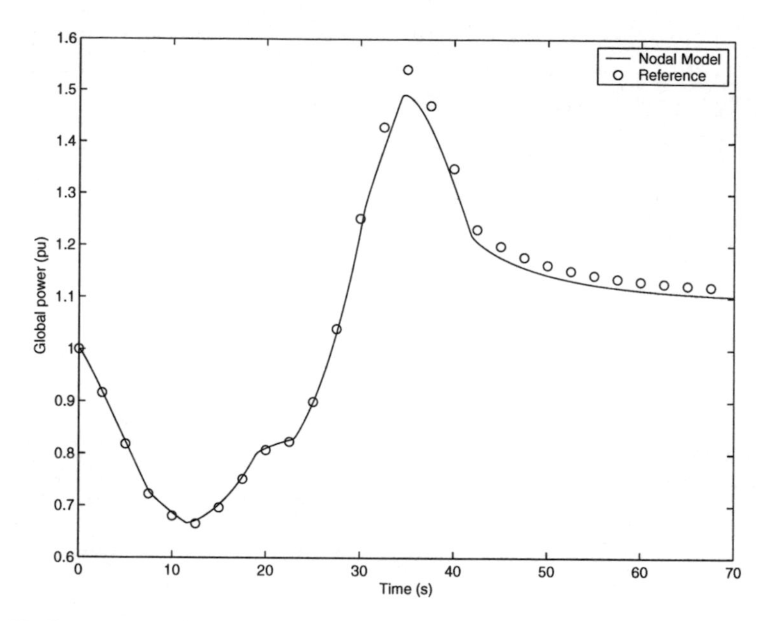

Fig. 2.20 Comparison of the Nodal Model Response with the Reference Solution during the Simultaneous Movement of All 8 RRs.

calculations carried out with a time step of 0.005 s. Fig. 2.19 shows a comparison of the nodal model response with the quasi–static reference response thus obtained, from which it is understood that the nodal model shows very good agreement with the reference response throughout.

Validation of the nodal model under a more severe total power transient involving simultaneous movement of all 8 RRs was also carried out. In this transient, all the RRs were considered to move simultaneously in the manner as depicted by Fig. 2.18. Fig. 2.20 shows the variation of total power obtained based on the nodal model. Reference values of total power are also shown. The total power reached up to about 150%FP which is far beyond the range of transients that may be encountered during normal reactor operation. Still, the accuracy of the nodal model is observed to be reasonable.

2.7.2 *Differential Movement of RRs*

Scenarios in which the power distribution in the reactor core undergoes variations in spite of the total power remaining constant are of great significance in spatial reactor control applications. In order to assess the validity of the nodal model under such conditions, a transient involving simultaneous counter–movement of two diagonally opposite RRs was simulated. The RR at lattice location $V9$ (in node 6) was driven linearly into the reactor core from its nominal position under a signal of 1V, until it reached the core bottom. RR at the diagonally opposite lattice locations $E17$ (in node 2) was driven out simultaneously at the same speed in order to maintain the net reactivity near to zero and thereby to keep the total power almost constant. Fig. 2.21 shows the position of these RRs during the transient. All other control rods were kept stationary during the transient. It is expected that during this transient, power in nodes belonging to one half of the reactor core containing the outgoing RR should increase, and that in the nodes belonging to the other half should decrease, such that the total power should remain more or less constant.

Instantaneous steady–state nodal power distribution calculated at different points of time using the finite difference model, with RR positions at their respective instantaneous values as in nodal model, was used as reference for validation of the nodal model response. Fig. 2.22 to Fig. 2.24 shows the variations in different nodal powers during the transient. Power in nodes 2, 3, 10, 11 and 17 increases significantly, whereas those in nodes 6, 7, 13, 14 and 15 decrease. Nodes 1, 4, 5, 8, 9, 12 and 16 lying on the interface of the above two diagonally opposite regions show less significant changes in power. It can again be noticed that the powers computed using the nodal model are in good agreement with the reference powers. Among all nodes, node 10 exhibits the maximum error of about 6.25%, towards the end of the transient when the RRs are at their maximum displaced position.

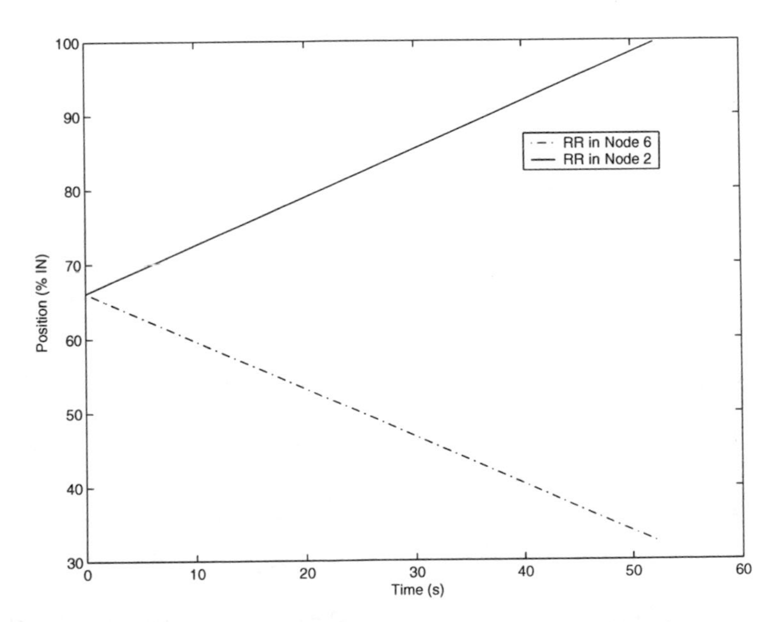

Fig. 2.21 Position of RRs in Nodes 2 and 6 during the Nodal Power Transient.

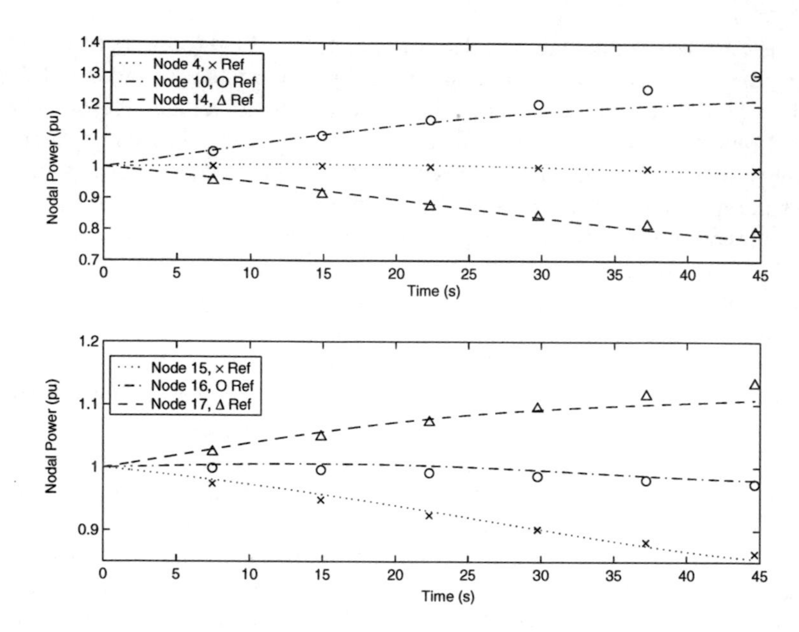

Fig. 2.22 Variations in Powers of Nodes 4, 10, 14 – 17 during Movement of Diagonally Opposite RRs.

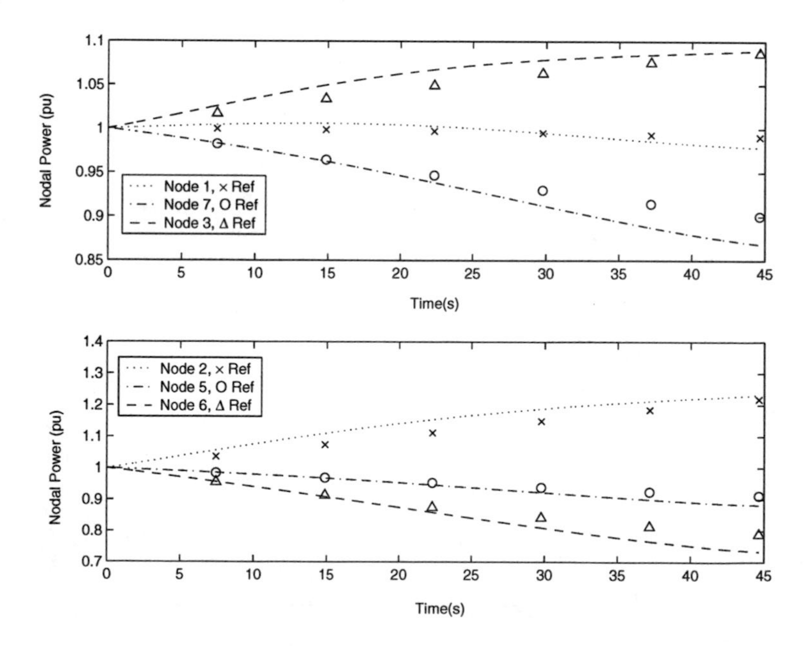

Fig. 2.23 Variations in Powers in Nodes 1 – 3 and 5 – 7 during Movement of Diagonally Opposite RRs.

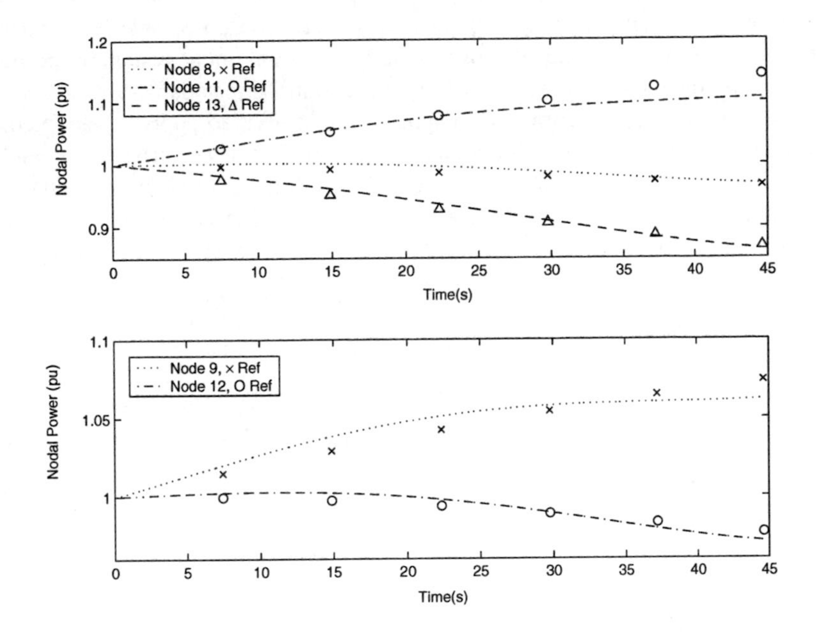

Fig. 2.24 Variations in Powers in Nodes 8, 9, 11 – 13 during Movement of Diagonally Opposite RRs.

2.8 Conclusion

Design and analysis of reactor and plant control systems require mathematical models. The models used for detailed core physics computations or burnup optimizations are not readily suited to control studies, which require lower order models. Among the various modeling methods, nodal, or multipoint kinetics model, is more suitable for control studies owing to its simplicity and structure of model equations is amenable to direct selection of state variables. In this chapter, an attempt has been made to develop a simple nodal model of the nuclear reactor, consisting of few first order ordinary differential equations. The approach for computation of coupling coefficients given here is much simpler than that based on adjoint operators and Greens functions. The final model equations are cast in a state space form suitable for control system studies. A suitable nodalization scheme and hence a model order can be arrived at by comparing the steady state accuracy and linear system properties such as stability, controllability and observability of various schemes. The model of the AHWR core neutronics thus arrived at is of much smaller order than the one used for core design.

It is observed that a scheme with 17 nodes would be appropriate for control related studies of AHWR, since it captures all the essential linear system properties as exhibited by detailed models, at the same time exhibiting satisfactory steady state and transient response accuracies. In spite of the simplified approach adopted, the model is found to be valid within a considerable region of perturbation around the steady state operating point, apart from capturing all the control relevant characteristics of a very fine model. Owing to the accuracy, simplicity and smaller order, this model is useful for simulation of control related transients. It can be efficiently used for controller design also.

Chapter 3
Output Feedback Control Design

3.1 Introduction

As of now, nuclear reactors all over the world employ output feedback controllers for control of total power and spatial power distribution. An output feedback controller can be designed through a conventional approach, *e.g.*, by plotting the root loci of the system. This is demonstrated in this chapter with application to the 17 node AHWR model.

The eigenvalues of the system matrix A of the AHWR model are given in Table 3.2, from which it can be noticed that the open loop eigenvalues of the AHWR model fall in three distinct clusters. First cluster has 38 eigenvalues ranging from -1.8870×10^{-4} to 7.4551×10^{-3}, second cluster of 35 eigenvalues, that ranges from -1.8395×10^{-1} to -1.2514×10^{-2} and the third one is of 17 eigenvalues ranging from -2.7626×10^{2} to -7.2516. This indicates that the AHWR exhibits a three–time–scale property. It can also be observed that the system has six eigenvalues with positive real parts besides the four eigenvalues at the origin.

The existence of multiple eigenvalues at origin and eigenvalues with positive real parts depicts instabilities in the AHWR. Hence it is necessary to devise a closed loop control that effectively maintains the total power of the reactor while the xenon induced oscillations are being controlled. Small and medium size nuclear reactors are generally controlled based on feedback of total power (or core averaged power), however large reactors may require feedback of spatial power distribution alongwith the total power feedback for effective spatial control. Here, characteristics of the closed loop system based on feedback of outputs has been investigated to find whether or not the reactor could be stabilized by output feedback [97].

3.2 Effect of Total Power Feedback on System Stability

At first the feasibility of stabilizing the system using the total power feedback alone is to be investigated. This can be easily achieved as follows.

S.R. Shimjith et al.: Modeling and Control of a Large Nuclear Reactor, LNCIS 431, pp. 61–77.
springerlink.com

It is assumed that the control signals to the RRs are generated based on the feedback of only the total reactor power *i.e.*, the input is given by

$$u_1 = -Ky, \tag{3.1}$$

in which

$$K = K_G = \begin{bmatrix} K_g\ 0\ 0 \ldots 0 \\ K_g\ 0\ 0 \ldots 0 \\ K_g\ 0\ 0 \ldots 0 \\ K_g\ 0\ 0 \ldots 0 \end{bmatrix}, \tag{3.2}$$

such that the feedback gain corresponding to total power is K_g for all RRs and is zero corresponding to nodal powers.

With (3.1), the system given by (2.75) becomes

$$\dot{z} = (A - B_1 K M)\, z + B_2 u_2 \tag{3.3}$$

$$y = Mz. \tag{3.4}$$

Now, the stability characteristics of the system for progressive increments in the value of K_g from zero can be investigated. It is observed that φ_1 and φ_6, the two unstable eigenvalues which were originally located at 7.4551×10^{-3} and 3.9654×10^{-6} respectively shift progressively towards the left side in the complex s–plane with increase in K_g. They reach -1.4715×10^{-1} and -3.712×10^{-5} respectively for $K_g \approx 20$ and remain there for any further increment in K_g, as shown in Fig. 3.1 and Fig. 3.2 respectively. Moreover, φ_7 and φ_{74}, the two eigenvalues which were originally at 0 and -7.251 respectively move towards each other with increment in K_g, until they reach a breakover point at -3.625 for $K_g = 421.1$. For $K_g > 421.1$, these eigenvalues break apart to form a complex conjugate pair of eigenvalues, as shown in Fig. 3.3. The effect of K_g on the location of the remaining 86 eigenvalues is not that significant barring for very slight shifts of few of them on the real axis, including the remaining three eigenvalues at origin which get shifted slightly to the left by about 10^{-8}.

With the consideration that an error of about 10% in the total power should lead to generation of full control signal ($\pm$1V), the value of total power control gain K_g is selected as 12.5. With this value of gain, the above mentioned eigenvalues which are sensitive to K_g are shifted to their respective new locations as listed in Table 3.1. Fig. 3.4 shows the closed loop transient response of the non–linear model with the total power feedback, for a change in total power setpoint from $90\%FP$ to $100\%FP$ at the rate of $0.5\%FP/s$. It can be observed that the total reactor power follows the demand power variations, with a small overshoot and steady state error. Although there are two unstable pairs of complex conjugate eigenvalues but their effect is not noticed in total power variation with time. This phenomenon, therefore, could be attributed to the presence of spatial instabilities.

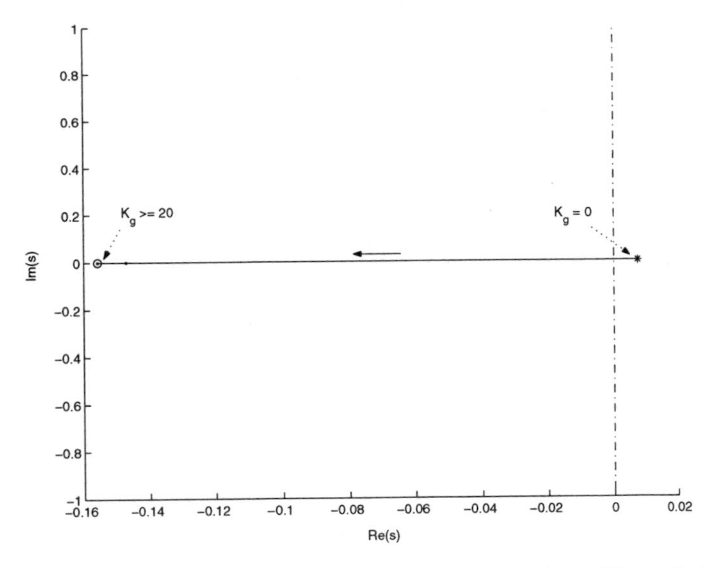

Fig. 3.1 Effect of Total Power Control Gain on Eigenvalue Locations: Change in location of φ_1 with increase in K_g.

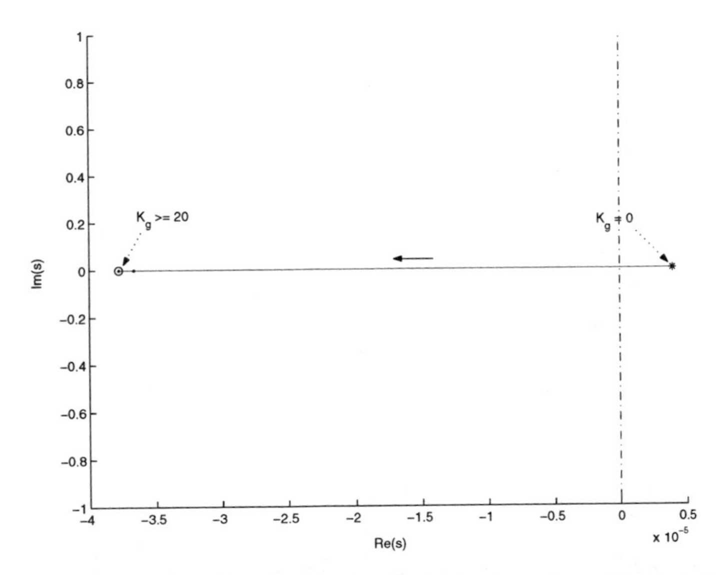

Fig. 3.2 Effect of Total Power Control Gain on Eigenvalue Locations: Change in location of φ_6 with increase in K_g.

To ascertain this, a transient involving a spatial power disturbance was simulated using the non–linear model. It was assumed that the reactor was initially operating at full power, with the control signal generated according to (3.1). A small disturbance was enforced in the position of the RR6 for a short duration. The RR,

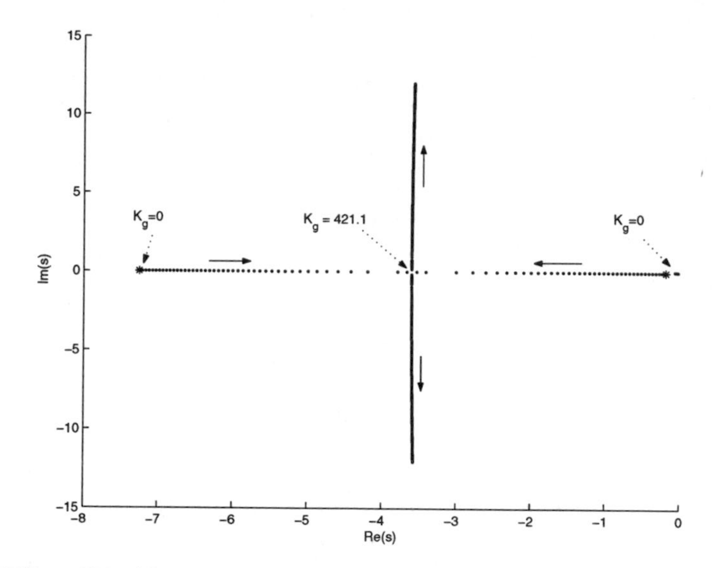

Fig. 3.3 Effect of Total Power Control Gain on Locations of φ_7 and φ_{74}.

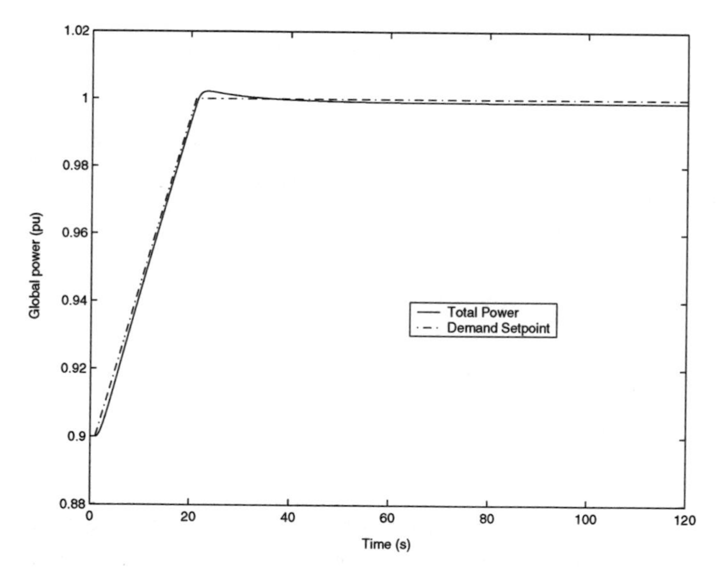

Fig. 3.4 Closed loop response of the system with total power feedback to a demand power change.

which was initially at its equilibrium position, was driven out by 2% under a control signal of $-1V$, and immediately driven in so as to come back to its original position under a signal of $+1V$. The response of the model, subsequent to this disturbance, was simulated for about 40 hours, and the characteristics of the system are

Table 3.1 Effect of Total Power Feedback.

	Open Loop Eigenvalue	Closed loop Eigenvalue (With $K_g = 12.5$)
φ_1	7.4551×10^{-3}	-4.8787×10^{-4}
φ_6	3.9654×10^{-6}	-2.6916×10^{-4}
φ_7	0	-4.0318×10^{-4}
φ_{74}	-7.2516	-7.2482

investigated in terms of the variations in total reactor power defined earlier and tilts in first azimuthal mode, second azimuthal mode and the first radial mode defined as

$$\text{First azimuthal tilt} = \frac{Q_L - Q_R}{Q} \times 100\%$$

$$\text{where } Q_L = \frac{1}{2}Q_1 + \sum_{h=6}^{9} Q_h + \sum_{h=14}^{17} Q_h,$$

$$\text{and } Q_R = \frac{1}{2}Q_1 + \sum_{h=2}^{5} Q_h + \sum_{h=10}^{13} Q_h;$$

$$\text{Second azimuthal tilt} = \frac{Q_{q1} - Q_{q2}}{Q} \times 100\%$$

$$\text{where } Q_{q1} = \frac{1}{2}Q_1 + Q_2 + Q_3 + Q_6 + Q_7 + Q_{10} + Q_{11} + Q_{14} + Q_{15},$$

$$\text{and } Q_{q2} = \frac{1}{2}Q_1 + Q_4 + Q_5 + Q_8 + Q_9 + Q_{12} + Q_{13} + Q_{16} + Q_{17}; \text{ and}$$

$$\text{First radial tilt} = \frac{Q_1 - \sum_{h=2}^{17} Q_h}{Q} \times 100\%.$$

It was observed in the simulation that the total reactor power is still being regulated at full power, but the power distribution in the core undergoes oscillations. First and second azimuthal modes of oscillations, respectively in terms of severity, are found to be dominant in AHWR. In the first azimuthal mode, powers in left and right halves of the reactor core oscillate in opposite phases. In the second azimuthal mode, sum of opposite quadrant powers oscillate out of phase with the sum of the other two quadrant powers. It was observed that within 40 hours, the first and second azimuthal modes of oscillations grow to amplitudes of the order of ±0.3% and ±0.2% respectively. In addition to these two modes, first radial mode of oscillations, eventhough much smaller in magnitude, is also observed to pick up. In first radial, power in the centre of the core (node 1) oscillates out of phase with the power in the rest of the core. It is noticed that this mode grows up to about ±0.03% within 40 hours of simulation. Fig. 3.5 depicts these tilts. The period of oscillations are observed to be about 12, 16 and 17 hours respectively for first azimuthal, second azimuthal and first radial modes. It should also be reckoned from the eigenvalues given

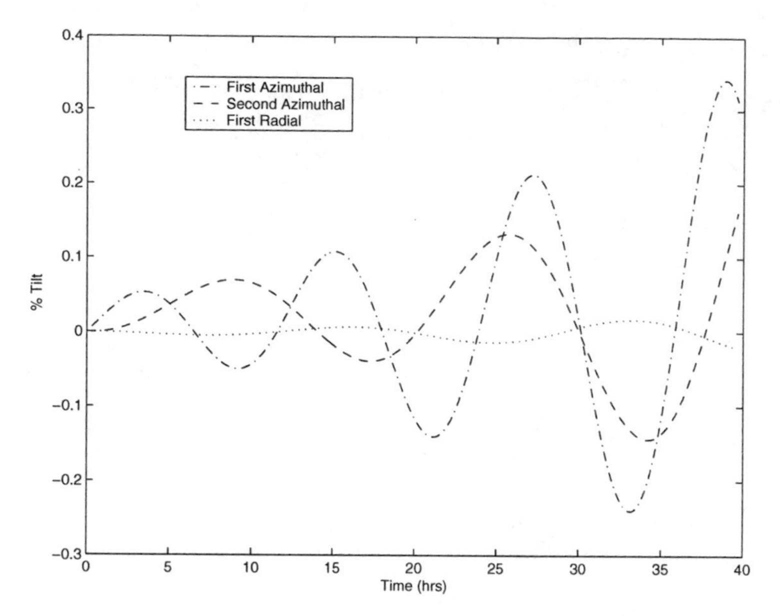

Fig. 3.5 Unstable Modes of Spatial Instability exhibited by Coupled Neutronics-Thermal Hydraulics Model of AHWR.

in Table 3.2 that the damped period of oscillation corresponding to the two unstable complex conjugate pairs of eigenvalues are 11.51 and 14.89 hours respectively.

If void reactivity feedback is not considered, the amplitude of spatial power oscillations are observed to be considerably higher for each of these modes, as depicted in Fig. 3.6. However, the frequency of oscillations and number of unstable modes remain more or less the same.

These spatial oscillations and subsequent local overpowers pose a potential threat to the fuel integrity of any nuclear reactor, and hence require control. Therefore it is necessary to devise a suitable spatial power controller for the AHWR.

3.3 Effect of Spatial Power Feedback on System Stability

As described in the previous section, it is evident that the spatial instabilities cannot be eradicated with feedback of only the total reactor power. Hence, in addition to the total power feedback, feedback of spatial power distribution must also be used for generation of the control signals to the RRs. Thus, in (3.1), K that was restricted to contain non–zero values in only first column, will now be allowed to have more non–zero values in other locations.

At first, spatial power feedback from two side halves of the reactor core are considered. First half consists of nodes $2,3,4,5,10,11,12,13$ and one half of node 1, and the other half of the core consists of the remaining nodes. It is assumed that

Table 3.2 Open-loop Eigenvalues of the AHWR Model with Thermal Hydraulics Feedback.

i	φ_i	i	φ_i	i	φ_i
1	7.4551×10^{-3}	39	-1.2514×10^{-2}	66	-1.6316×10^{-1}
$2-3$	$(8.8268 \pm j1.8656) \times 10^{-5}$	40	-1.6108×10^{-2}	67	-1.6325×10^{-1}
$4-5$	$(8.0470 \pm j2.4129) \times 10^{-5}$	41	-5.0954×10^{-2}	68	-1.6405×10^{-1}
6	3.9654×10^{-6}	42	-5.1159×10^{-2}	69	-1.6576×10^{-1}
$7-10$	0	43	-5.7730×10^{-2}	70	-1.8037×10^{-1}
$11-12$	$(-3.5182 \pm j7.7577) \times 10^{-5}$	44	-5.7893×10^{-2}	71	-1.8049×10^{-1}
13	-3.7781×10^{-5}	45	-5.9707×10^{-2}	72	-1.8122×10^{-1}
$14-15$	$(-3.7785 \pm j7.6475) \times 10^{-5}$	46	-5.9723×10^{-2}	73	-1.8395×10^{-1}
16	-3.7993×10^{-5}	47	-6.0344×10^{-2}	74	-7.2516
17	-4.0124×10^{-5}	48	-6.0642×10^{-2}	75	-3.2844×10^{1}
18	-4.1520×10^{-5}	49	-6.1848×10^{-2}	76	-3.3372×10^{1}
19	-4.2245×10^{-5}	50	-6.1942×10^{-2}	77	-6.6599×10^{1}
20	-4.4204×10^{-5}	51	-6.2200×10^{-2}	78	-6.8323×10^{1}
21	-4.7476×10^{-5}	52	-6.2380×10^{-2}	79	-9.3653×10^{1}
22	-4.8866×10^{-5}	53	-6.2458×10^{-2}	80	-9.4612×10^{1}
$23-24$	$(-6.4855 \pm j5.3109) \times 10^{-5}$	54	-6.2608×10^{-2}	81	-1.0868×10^{2}
$25-26$	$(-6.5890 \pm j5.4696) \times 10^{-5}$	55	-6.2865×10^{-2}	82	-1.1705×10^{2}
$27-28$	$(-7.3359 \pm j3.9319) \times 10^{-5}$	56	-6.2893×10^{-2}	83	-1.6967×10^{2}
$29-30$	$(-7.7407 \pm j2.9929) \times 10^{-5}$	57	-1.1714×10^{-1}	84	-1.7568×10^{2}
31	-1.4107×10^{-4}	58	-1.4712×10^{-1}	85	-1.9497×10^{2}
32	-1.4624×10^{-4}	59	-1.4713×10^{-1}	86	-2.1110×10^{2}
33	-1.5717×10^{-4}	60	-1.4809×10^{-1}	87	-2.1904×10^{2}
34	-1.6524×10^{-4}	61	-1.4849×10^{-1}	88	-2.3591×10^{2}
35	-1.6720×10^{-4}	62	-1.5580×10^{-1}	89	-2.7163×10^{2}
36	-1.7308×10^{-4}	63	-1.5585×10^{-1}	90	-2.7626×10^{2}
37	-1.8807×10^{-4}	64	-1.5662×10^{-1}		
38	-1.8870×10^{-4}	65	-1.5759×10^{-1}		

the spatial control signals to the RRs are generated using the feedback of power generated in these two radial halves of the reactor core, *i.e.*, in (3.1), $K = K_S$ where

$$K_S = \begin{bmatrix} K_g & K_s/2 & K_s & K_s & K_s & K_s & 0 & 0 & 0 & 0 & K_s & K_s & K_s & K_s & 0 & 0 & 0 & 0 \\ K_g & K_s/2 & K_s & K_s & K_s & K_s & 0 & 0 & 0 & 0 & K_s & K_s & K_s & K_s & 0 & 0 & 0 & 0 \\ K_g & K_s/2 & 0 & 0 & 0 & 0 & K_s & K_s & K_s & K_s & 0 & 0 & 0 & 0 & K_s & K_s & K_s & K_s \\ K_g & K_s/2 & 0 & 0 & 0 & 0 & K_s & K_s & K_s & K_s & 0 & 0 & 0 & 0 & K_s & K_s & K_s & K_s \end{bmatrix} \quad (3.5)$$

Now the shift in eigenvalues of the closed loop system (3.3) is considered for gradually increasing values of gain K_s from zero while K_g is fixed at 12.5. Fig. 3.7 shows the locus of both the stable and unstable eigenvalues near the origin. A complex conjugate pair of eigenvalues originally located at $(8.8268 \pm j1.8656) \times 10^{-5}$ shifts in the left half of complex s–plane for $K_s \geq 0.005$. However, the pair of unstable eigenvalues located at $(8.0470 \pm j2.4129) \times 10^{-5}$ is insensitive to the variations in

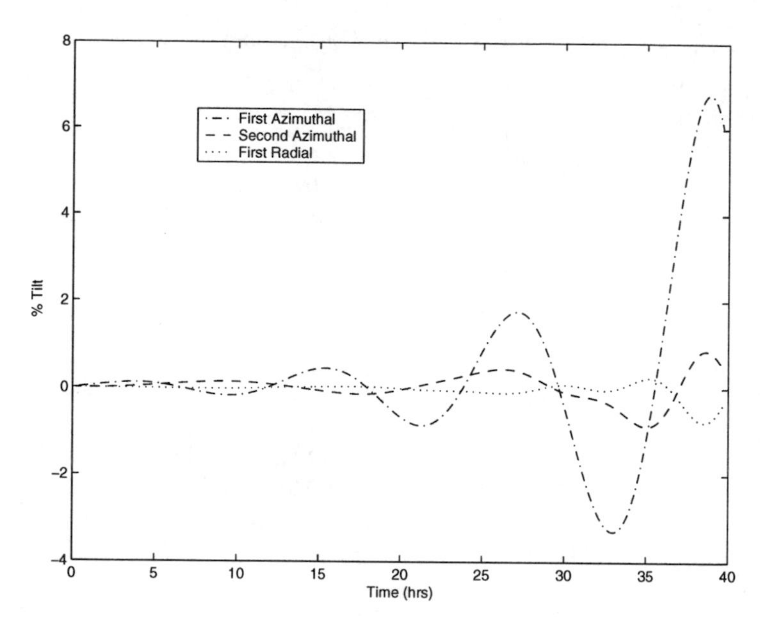

Fig. 3.6 Unstable Modes of Spatial Instability in the Absence of Thermal Hydraulics Feedback.

K_s. This implies that complete spatial stabilization of the AHWR cannot be achieved with power feedback from only two halves of the reactor core.

Subsequently, spatial stabilization with power feedback from four quadrants of the reactor core is attempted. The spatial control signal to the RR in each quadrant of the core is assumed to be generated based on the deviation of the power generated in that quadrant from the setpoint. This is effected if $K = K_Q$ where

$$K_Q = \begin{bmatrix} K_g & K_q/4 & K_q & K_q & 0 & 0 & 0 & 0 & 0 & 0 & K_q & K_q & 0 & 0 & 0 & 0 & 0 & 0 \\ K_g & K_q/4 & 0 & 0 & K_q & K_q & 0 & 0 & 0 & 0 & 0 & 0 & K_q & K_q & 0 & 0 & 0 & 0 \\ K_g & K_q/4 & 0 & 0 & 0 & 0 & K_q & K_q & 0 & 0 & 0 & 0 & 0 & 0 & K_q & K_q & 0 & 0 \\ K_g & K_q/4 & 0 & 0 & 0 & 0 & 0 & 0 & K_q & K_q & 0 & 0 & 0 & 0 & 0 & 0 & K_q & K_q \end{bmatrix}. \quad (3.6)$$

Fig. 3.8 shows the locus of both the stable and unstable eigenvalues near the origin when K_q was increased progressively from zero. It can be observed that all the unstable eigenvalues are stabilized for $K_q \geq 1.2$, which confirms that the quadrant core power feedback can be effectively used to completely stabilize the system. This corroborates the control philosophy of AHWR proposed in [21].

Most of the eigenvalues near the origin are found to have settled at their respective new locations for $K_q \approx 10$. With this consideration, value of spatial power control gain K_q is selected as 10. Table 3.3 lists the closed loop eigenvalues with this spatial power controller alongwith the total power control gain of 12.5.

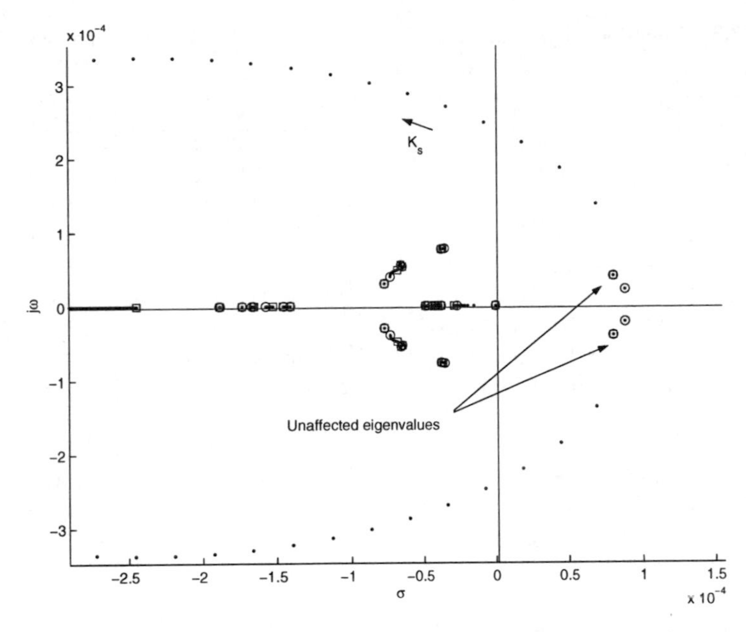

Fig. 3.7 Effect of Spatial Power Feedback from Two Halves of the Reactor Core.

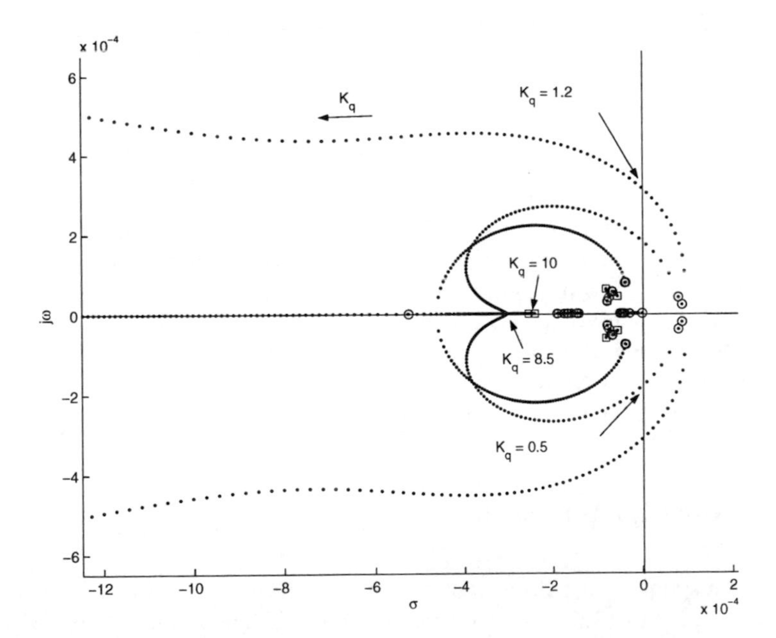

Fig. 3.8 Effect of Spatial Power Feedback from Four Quadrants of the Reactor Core.

Table 3.3 Closed-loop Eigenvalues of the AHWR Model with Total and Spatial Power Controls.

i	φ_i	i	φ_i	i	φ_i
1	-2.8438×10^{-5}	43	-5.7657×10^{-2}	75	-3.2840×10^{1}
2	-2.8440×10^{-5}	44	-5.7733×10^{-2}	76	-3.3368×10^{1}
3	-2.8441×10^{-5}	45	-5.9707×10^{-2}	77	-6.6599×10^{1}
4	-2.8487×10^{-5}	46	-5.9708×10^{-2}	78	-6.8321×10^{1}
$5-6$	$(-3.6346\pm+j7.6958)\times10^{-5}$	47	-6.0332×10^{-2}	79	-9.3653×10^{1}
7	-3.8196×10^{-5}	48	-6.0641×10^{-2}	80	-9.4612×10^{1}
$8-9$	$-3.9700\times10^{-5}\pm j3.8154\times10^{-7}$	49	-6.1850×10^{-2}	81	-1.0868×10^{2}
$10-11$	$-4.2124\times10^{-5}\pm j3.8165\times10^{-8}$	50	-6.1943×10^{-2}	82	-1.1705×10^{2}
12	-4.3957×10^{-5}	51	-6.2200×10^{-2}	83	-1.6967×10^{2}
13	-4.6923×10^{-5}	52	-6.2380×10^{-2}	84	-1.7568×10^{2}
14	-4.7854×10^{-5}	53	-6.2457×10^{-2}	85	-1.9497×10^{2}
$15-16$	$(-5.4349\pm j4.3103)\times10^{-5}$	54	-6.2608×10^{-2}	86	-2.1110×10^{2}
$17-18$	$(-6.6416\pm j5.4612)\times10^{-5}$	55	-6.2864×10^{-2}	87	-2.1904×10^{2}
$19-20$	$(-7.6987\pm j3.0881)\times10^{-5}$	56	-6.2893×10^{-2}	88	-2.3591×10^{2}
$21-22$	$(-8.0238\pm j6.0497)\times10^{-5}$	57	-1.1720×10^{-1}	89	-2.7163×10^{2}
23	-1.4139×10^{-4}	58	-1.4712×10^{-1}	90	-2.7626×10^{2}
24	-1.4564×10^{-4}	59	-1.4714×10^{-1}		
25	-1.5219×10^{-4}	60	-1.4811×10^{-1}		
26	-1.5848×10^{-4}	61	-1.4856×10^{-1}		
27	-1.6325×10^{-4}	62	-1.5581×10^{-1}		
28	-1.7360×10^{-4}	63	-1.5586×10^{-1}		
29	-1.8017×10^{-4}	64	-1.5665×10^{-1}		
30	-1.8619×10^{-4}	65	-1.5773×10^{-1}		
31	-2.5643×10^{-4}	66	-1.6317×10^{-1}		
32	-2.5862×10^{-4}	67	-1.6326×10^{-1}		
33	-2.7265×10^{-4}	68	-1.6409×10^{-1}		
34	-2.7313×10^{-4}	69	-1.6608×10^{-1}		
35	-2.0853×10^{-3}	70	-1.8038×10^{-1}		
$36-37$	$(-4.7046\pm j1.4643)\times10^{-3}$	71	-1.8051×10^{-1}		
$38-39$	$(-1.4443\pm j3.4673)\times10^{-2}$	72	-1.8125×10^{-1}		
40	-1.5760×10^{-2}	73	-1.8450×10^{-1}		
$41-42$	$-4.9956\times10^{-2}\pm j4.2559\times10^{-4}$	74	-7.2266		

3.4 Closed Loop Response

Transient response of the closed loop system with the control law (3.1), with the gain matrix specified by (3.6), needs to be assessed through non–linear simulations. The reactor is assumed to be initially operating at full power equilibrium conditions. Shortly, RR2, originally under auto control, is manually driven out by about 2%

under a signal of $-1V$, and left on auto control afterwards. RR4, RR6 and RR8 are under auto control and total and spatial power controllers were active throughout. Other four RRs which are under manual control are assumed to remain stationary throughout the simulation.

It may be noted that one of the basic assumptions in development of (2.59) and (2.60) is that the steam drum level is being maintained, for which it is essential that the steam flow and feed flow should be equal and proportional to the total reactor power. Therefore throughout this transient, the feed flow rate q_f was assumed to be directly proportional to the total reactor power as

$$q_f = q_{f0} \frac{\sum_h Q_h}{\sum_h Q_{h0}}, \tag{3.7}$$

where q_{f0} is the feed water flow rate at full power operating conditions.

During the manual retraction of RR2, it is observed that the total power experienced an increase by about 0.3%. However, subsequently the power stabilized back at the rated value, as shown in Fig. 3.9. Positions of all RRs are depicted in Fig. 3.10. It may be noticed that while RR2 is being driven out manually, other RRs moves in under the effect of the controller in order to maintain the total power. After the period during which the manual signal was enforced on RR2, all the RRs are being driven back by the controller to their original positions. Fig. 3.11 depicts the variations in normalized quadrant core powers, from which it may be observed that the power in the first quadrant which contains RR2 experienced an increase by about 0.1%, and it is being compensated by decrement in power generated in other quadrants of the core. The disturbances in the quadrant core powers are suppressed by the controller within about 150 seconds. It can be observed from Fig. 3.12 that the perturbations in nodal powers are also being eliminated by the controller. Furthermore, on continuing the simulation over an extended period, it was observed that the power tilts are not getting surfaced.

Fig. 3.13 shows the closed loop system response during another transient initiated by a momentary disturbance in position of RR6. The simulation was carried out with the non–linear model with a time step of 5 ms and data logging interval of 5 minutes. It was assumed that the reactor was initially operating at full power, with the control signal generated according to (3.1). The RR6, which was initially at its equilibrium position, was driven out by about 1.5% under a control signal of $-1V$, and immediately driven in so as to come back to its original position under a signal of $+1V$. As a consequence of this disturbance, the tilts started picking up. After about 200 minutes, the spatial control component was introduced in the control signals by switching the control law to that described by (3.6). Fig. 3.13(a) shows the variations in tilts and Fig. 3.13(b) shows a zoomed version of the former, focusing the region near the introduction of spatial control. It was observed that the tilts were suppressed within about 5 minutes.

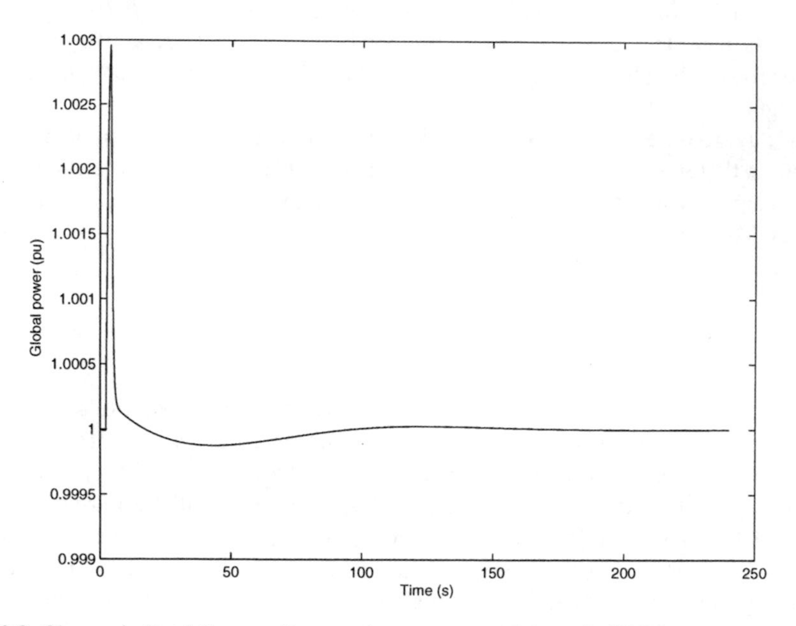

Fig. 3.9 Change in Total Reactor Power subsequent to withdrawal of RR2.

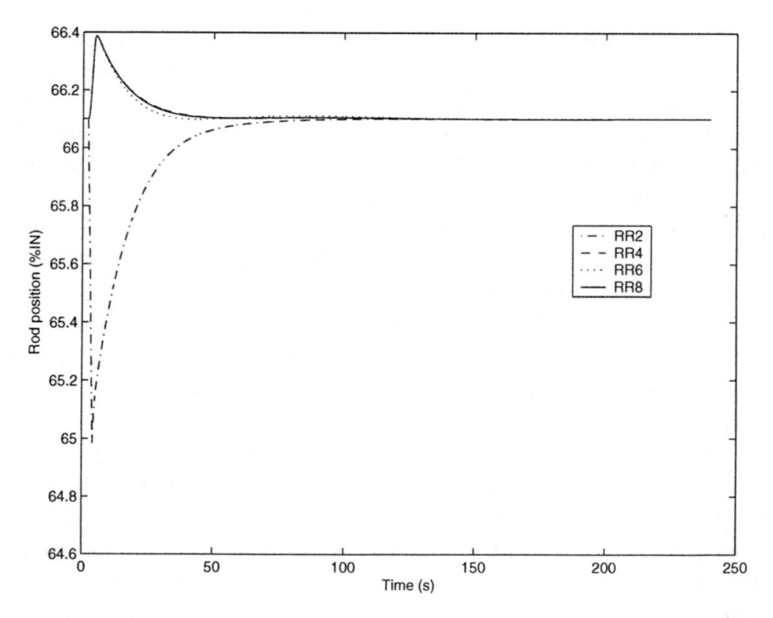

Fig. 3.10 RR Positions During the Transient.

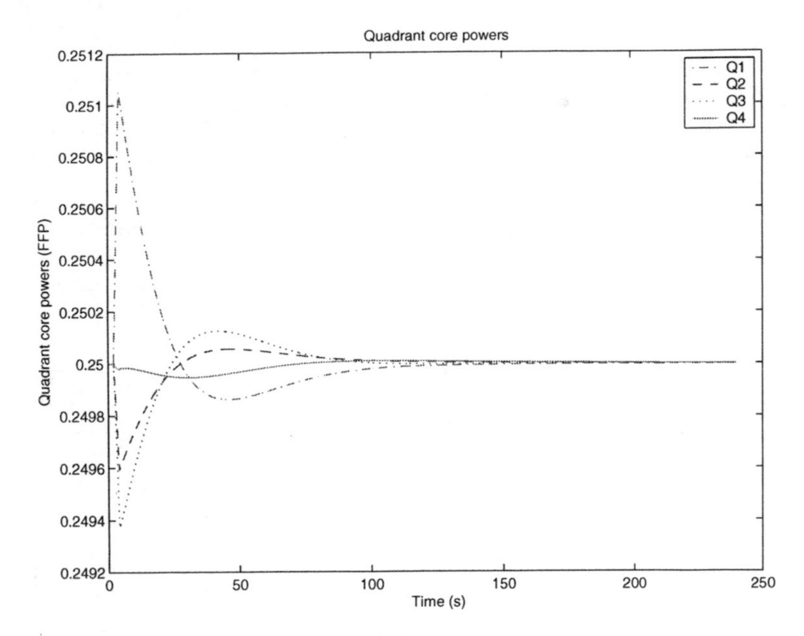

Fig. 3.11 Quadrant Core Powers during the Transient.

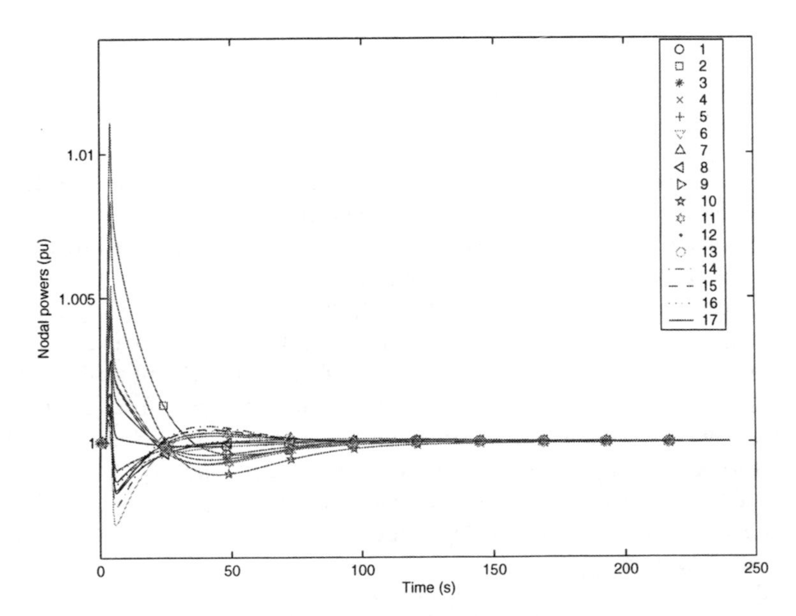

Fig. 3.12 Changes in Nodal Powers following Withdrawal of RR2.

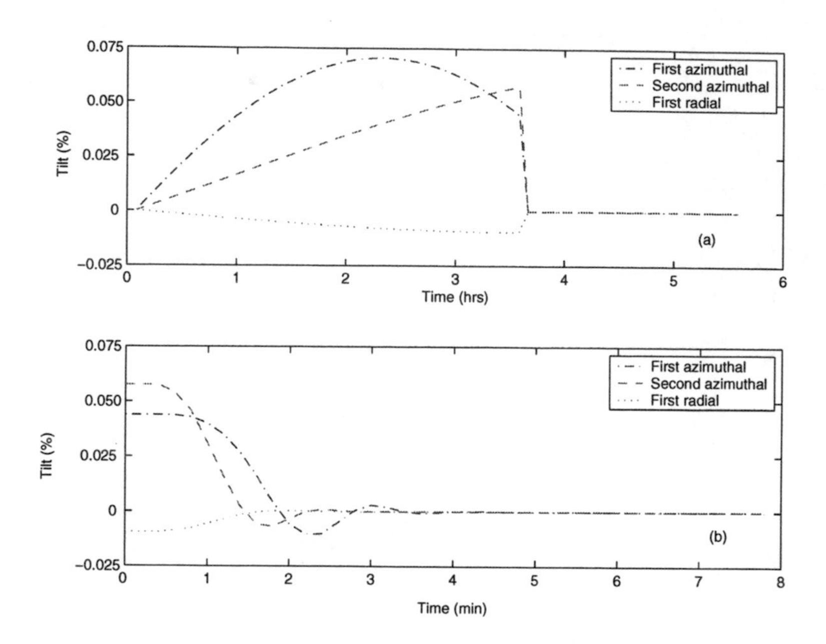

Fig. 3.13 Suppression of Tilts.

3.4.1 Robustness of the Controller

Robustness issues are not explicitely taken into consideration during the design of the above mentioned output feedback controller. However, any changes in feed flow rate q_f from that given by (3.7) acts as a disturbance on the system. In order to assess the response of the system to disturbances in feed flow, a scenario was simulated in which the reactor was operating at steady full power when a 5% positive step change was introduced in the feed flow as shown in Fig. 3.14. As a result of this, the incoming coolant enthalpy reduced by about 0.64% as shown in Fig. 3.15 and exit qualities from each node underwent changes as shown in Fig. 3.16. Total as well as nodal powers were found to be stabilizing at power levels of about 1.6% higher than the respective original values, as shown in Fig. 3.17 and Fig. 3.18. This could be justified by (2.60) and (2.53) that in order for the incoming enthalpy to stabilize after an increase in feed flow, the nodal powers Q_h and subsequently the core flows q_{d_h} should also increase.

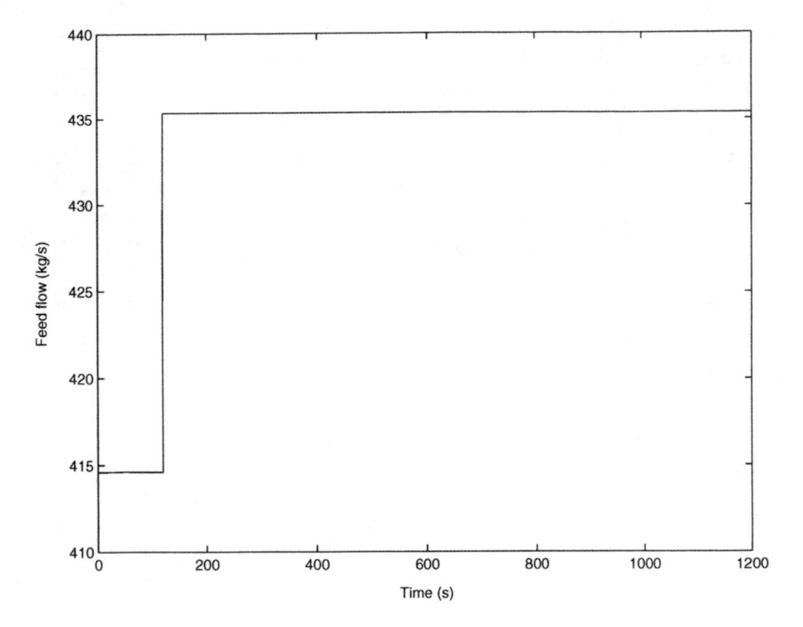

Fig. 3.14 Step change in the feed flow.

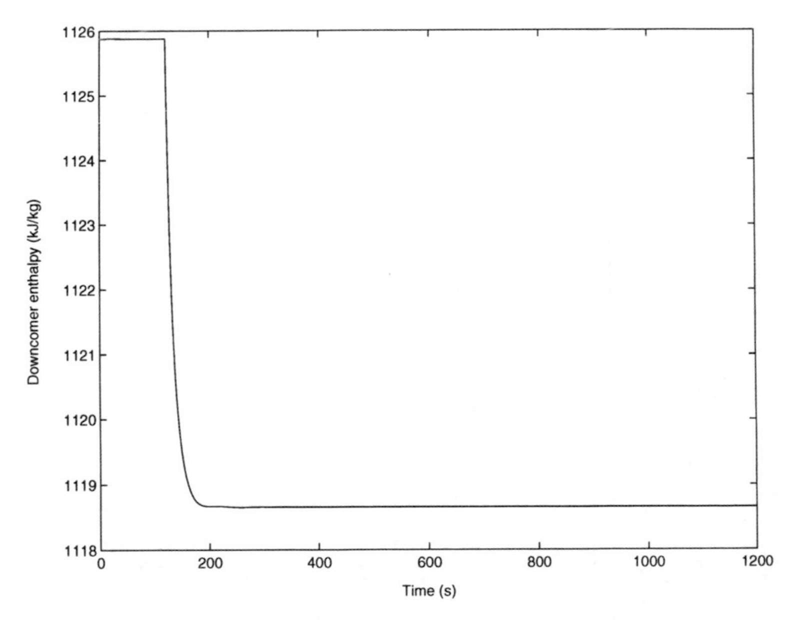

Fig. 3.15 Change in incoming coolant enthalpy due to step change in the feed flow.

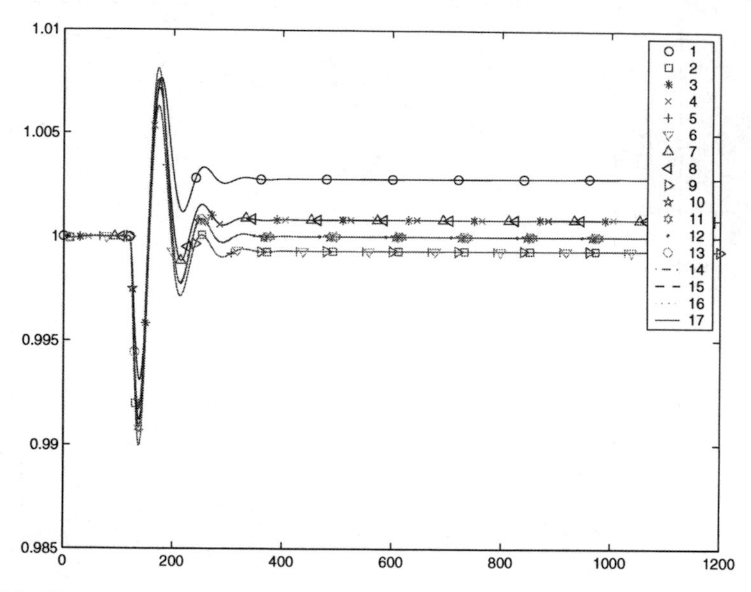

Fig. 3.16 Changes in exit qualities due to step change in the feed flow.

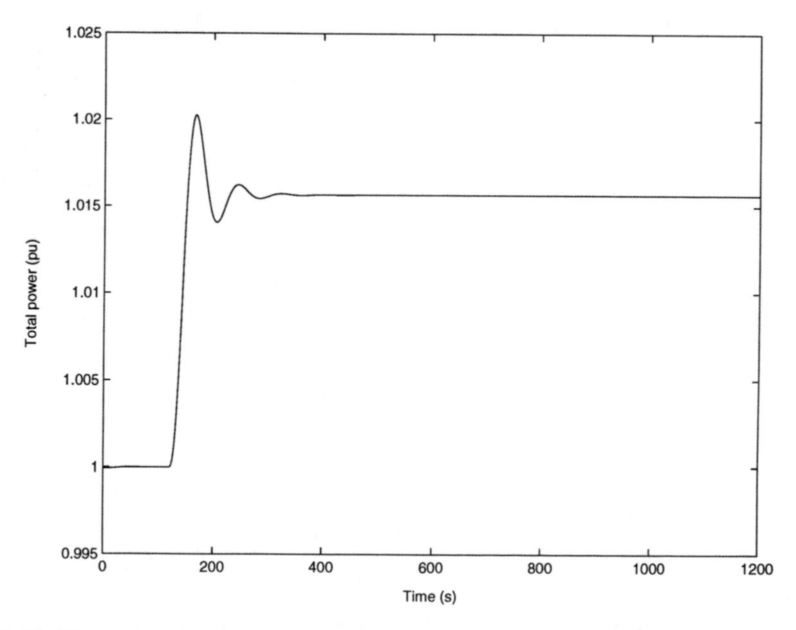

Fig. 3.17 Change in total reactor power due to step change in the feed flow.

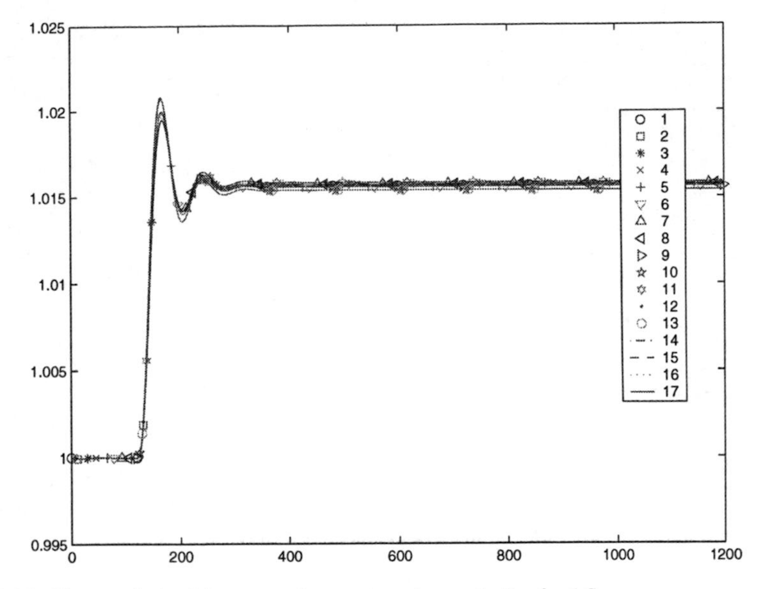

Fig. 3.18 Changes in nodal powers due to step change in the feed flow.

3.5 Conclusion

The Advanced Heavy Water Reactor exhibits xenon induced spatial power instability due to the large core size and loose neutron coupling. Three modes of oscillations are found to be dominant in AHWR. Reactivity feedback due to voids contributes slightly to system stability, by introducing some damping on the oscillatory response. Analysis of effect of total power feedback gain on system stability proves that a control strategy based on feedback of only the total reactor power cannot eradicate the spatial instabilities. It is further established that complete system stability can be achieved by introducing a spatial power control component based on feedback of quadrant core powers, along with the total power control component. The efficacy and robustness of the conventional controller thus designed is demonstrated through dynamic simulations. It is observed that the controller stabilizes total as well as spatial powers effectively even under disturbances.

Chapter 4
Multiparameter Singular Perturbation of Linear Optimal Regulators

4.1 Introduction

In a nuclear reactor, the fission of ^{235}U, on an average, is characterized by the release of two to three neutrons which fall under two categories namely prompt and delayed neutrons[24]. The former are released almost instantaneously whereas the latter are released after significant time delays subsequent to a fission. The time delays associated with delayed neutrons vary from a fraction of a second to about a minute. Moreover, phenomena due to fission products like xenon have even much larger time constants, of the order of several hours. Hence, a nuclear reactor system is essentially characterized by a multi–time–scale behavior. Likewise the AHWR also exhibits a three–time–scale behavior with eigenvalues falling in three distinct and widely separated clusters, as described in chapter 4.

Multitime–scale behavior makes a system model susceptible to numerical ill–conditioning. Decomposition of such models into two subsystems, based on singular perturbation methods, have already been applied to the nuclear reactor control problems to handle the ill–conditioning effects[88, 114]. Singularly perturbed systems, more generally multitime scale systems, often occur naturally in mathematical models due to the presence of small constants, masses, large feedback gains, weak coupling *etc*. It was recognized long ago that the singular perturbations are present in most classical and modern control schemes based on reduced order models, and it led to the development of time-scale methods for a variety of applications including state feedback, output feedback, filter and observer design. First survey of control theory applications of singular perturbations appeared in 1976 due to Kokotovic[56]. Subsequently a large number of books and surveys came up in this area [4, 23, 25, 68, 73, 91].

Majority of the literature in singular perturbations deals with two–time scale systems, which in general, can be modeled by the set of nonlinear differential equations as

$$\dot{z_1} = f(z_1, z_2, t), \;\; z_1(t_0) = z_1^0,$$

S.R. Shimjith et al.: Modeling and Control of a Large Nuclear Reactor, LNCIS 431, pp. 79–100.
springerlink.com

$$\dot{z_2} = G(z_1, z_2, t), \ \ z_2(t_0) = z_2^0 \tag{4.1}$$

where the n-dimensional vector z_1 is predominantly slow and the m-vector z_2 contains fast transients superimposed on a slowly varying 'quasi-steady state', that is $||\dot{z_2}|| >> ||\dot{z_1}||$. This is quite often expressed by introducing $g = \varepsilon G$ thereby scaling g to be of the same order of magnitude as f. Scaling parameter ε is the speed ratio of the slow versus fast phenomena and multiplies the derivative of z. Thus,

$$\begin{aligned} \dot{z_1} &= f(z_1, z_2, t), \ \ z_1(t_0) = z_1^0 \\ \varepsilon \dot{z_2} &= g(z_1, z_2, t), \ \ z_2(t_0) = z_2^0. \end{aligned} \tag{4.2}$$

A linear time invariant version of (5.2) takes the form

$$\begin{bmatrix} \dot{z_1} \\ \dot{z_2} \end{bmatrix} = \begin{bmatrix} A & B \\ \hat{C} & \hat{D} \end{bmatrix} \begin{bmatrix} z_1 \\ z_2 \end{bmatrix}. \tag{4.3}$$

Separation of states into slow and fast is a nontrivial modeling task demanding ingenuity and insight[54], [55]. On applying the transformation $\eta = z_2 + Lz_1$ [55] where L satisfies the algebraic Riccati equation

$$\hat{D}L - LA + LBL - \hat{C} = 0,$$

(4.3) can be transformed into the block triangular system

$$\begin{bmatrix} \dot{z_1} \\ \dot{\eta} \end{bmatrix} = \begin{bmatrix} F_1 & B \\ 0 & F_2 \end{bmatrix} \begin{bmatrix} z_1 \\ \eta \end{bmatrix} \tag{4.4}$$

where $F_1 = A - BL$, $F_2 = \hat{D} + LB$ and the largest eigenvalue of F_1 is smaller than the smallest eigenvalue of F_2. If the eigenvalue separation of F_1 and F_2 are sufficiently large, complete separation of fast and slow subsystems are obtained by permitting $\varsigma = z_1 - M\eta$ where M satisfies

$$(A - BL)M - M\left(\hat{D} + LB\right) + B = 0$$

which results in

$$\dot{\varsigma} = (A - BL)\varsigma \tag{4.5}$$

$$\dot{\eta} = \left(\hat{D} + LB\right)\eta \tag{4.6}$$

where ς are the slow states and η are the fast states.

In order to guarantee the stability of the original linear singularly perturbed system, it is sufficient to guarantee the stability of the decomposed subsystems[45]. Likewise, for a linear system decomposed into slow and fast subsystems as

$$\begin{aligned} \dot{z}_s &= A_0 z_s + B_0 u_s, \\ y_s &= C_0 z_s + D_0 u_s, \end{aligned}$$

$$\varepsilon \dot{z}_f = A_1 z_f + B_1 u_f,$$
$$y_f = C_1 z_f,$$

observability and controllability of original system follows respectively from observability and controllability of subsystems[77].

Decomposition of two–time scale systems into separate slow and fast subsystems suggests that separate slow and fast control laws can be designed for each subsystem and then combined into a composite control of the original system. Proof for the near optimality of the composite controller for a quadratic cost function can be found, *e.g.*, in [18].

If within F_1 and F_2 eigenvalues fall within distinct clusters, then (4.4) is called a multi–time scale system. Coupled with high order and multi–time–scale behavior, such system models may defy the direct application of two–time–scale methods. One way of decomposing such a system is by sequential application of two–time scale decomposition[101]. An alternate approach for the study of singularly perturbed linear system with multiparameters and multi–time scales is given by Ladde *et.al.*[58, 59]. Consider a set of linear differential equations described by

$$\begin{aligned}
\dot{z_1} &= A_0(t) z_1 + \sum_{k=1}^{r} B_{0k}(t) z_{2_k} + \sum_{k=1}^{s} C_{0k}(t) z_{3_k} \\
\varepsilon_i \dot{z}_{2i} &= A_{i0}(t) z_1 + \sum_{k=1}^{r} B_{ik}(t) z_{2_k} + \sum_{k=1}^{s} C_{ik}(t) z_{3_k} \\
\mu_j \dot{z}_{3_j} &= A_{j0}(t) z_1 + \sum_{k=1}^{r} B_{jk}(t) z_{2_k} + \sum_{k=1}^{s} C_{jk}(t) z_{3_k}
\end{aligned} \tag{4.7}$$

where $z_1 \in \Re^{n_0}$, $z_{2_i} \in \Re^{m_i}$, $z_{3_j} \in \Re^{l_j}$ and the dimensions of the entire system is $n = n_0 + m + l$ where $m = \sum_i m_i$ and $l = \sum_j l_j$. Parameters ε_i and μ_j are positive bounded real numbers of same order, separately. Assuming that ε_i's and μ_j's are not of the same order and defining

$$\varepsilon = (\varepsilon_1 \varepsilon_2 \ldots \varepsilon_r)^{1/r}, \quad \mu = (\mu_1 \mu_2 \ldots \mu_s)^{1/s},$$

it can be seen that (4.7) is a three time scale multiparameter system given by

$$\begin{aligned}
\dot{z_1} &= A_0(t) z_1 + A_{01}(t) z_2 + A_{02}(t) z_3 \\
\varepsilon \dot{z_2} &= D_1 A_{10}(t) z_1 + D_1 A_{11}(t) z_2 + D_1 A_{12}(t) z_3 \\
\mu \dot{z_3} &= D_2 A_{20}(t) z_1 + D_2 A_{21}(t) z_2 + D_2 A_{22}(t) z_3
\end{aligned} \tag{4.8}$$

where $z_2 = (z_{2_1}^T, z_{2_2}^T, \ldots, z_{2_r}^T)^T$, $z_3 = (z_{3_1}^T, z_{3_2}^T, \ldots, z_{3_s}^T)^T$ and D_i are diagonal matrices of suitable order with $\varepsilon/\varepsilon_i$'s and μ/μ_j's as entries.

Ladde *et.al.* [58] carried out a two stage reduction of (4.8) by setting $\varepsilon = 0$ and $\mu = 0$ respectively, to result in an overall reduced system, ε reduced system and μ reduced systems in respective ‘stretched’ time scales as

$$\dot{\tilde{z}}_1 = \Big(A_0 - A_{02}A_{22}^{-1}A_{20} - \big(A_{01} - A_{02}A_{22}^{-1}A_{21}\big)\big(A_{11} - A_{12}A_{22}^{-1}A_{21}\big)^{-1}\big(A_{10} - A_{12}A_{22}^{-1}A_{20}\big)\Big)\tilde{z}_1,$$
$$\dot{\hat{z}}_2 = D_1\left(A_{11} - A_{12}A_{22}^{-1}A_{21}\right)\hat{z}_2,$$
$$\dot{\hat{z}}_3 = D_2 A_{22}\hat{z}_3. \tag{4.9}$$

They also developed a three–fold version of Chang's transformation [15] which directly results (4.8) into a totally decoupled diagonal form and derived iterative methods for computation of the coefficients of the necessary similarity transformation matrix. However, they focused their study only to systems with no input or output matrices. Similar results in decomposition of three time scale autonomous systems, with more extensive mathematical background, can also be found in Gaitsgory *et.al.* [29].

As an extension to these, here the simultaneous decomposition of a non–autonomous singularly perturbed system into three subsystems namely 'slow', 'fast 1' and 'fast 2' respectively is considered. Design of a composite controller in terms of the individual subsystem controllers is derived, and the decomposition of the optimal control problem of the original high order system into three smaller order optimal control problems is discussed. The efficacy of the model decomposition and control design methods proposed here has been demonstrated by applying to the spatial control problem of AHWR.

4.2 System Decomposition

Consider that the model of the system is available in the following form:

$$\dot{z_1} = A_{11}z_1 + A_{12}z_2 + A_{13}z_3 + B_1u, \quad z_1(0) = z_{1_0} \tag{4.10}$$
$$\varepsilon\dot{z_2} = A_{21}z_1 + A_{22}z_2 + A_{23}z_3 + B_2u, \quad z_2(0) = z_{2_0} \tag{4.11}$$
$$\mu\dot{z_3} = A_{31}z_1 + A_{32}z_2 + A_{33}z_3 + B_3u, \quad z_3(0) = z_{3_0} \tag{4.12}$$
$$y = M_1z_1 + M_2z_2 + M_3z_3. \tag{4.13}$$

where the n_1 dimensional state vector z_1 denotes the set of predominantly slow varying states, n_2 dimensional state vector z_2 contains fast transients superimposed on a slowly varying quasi–steady–state and the n_3 dimensional state vector z_3 contains very fast transients superimposed on a slowly varying quasi–steady–state. u denotes m dimensional input vector and y denotes the p dimensional output vector. A_{ij}, B_i and M_i are matrices of appropriate dimensions. ε and μ which respectively represent the speed ratios of fast and very fast versus slow phenomena, are small positive parameters, such that $\mu \ll \varepsilon$.

Setting $\mu = 0$ we get from (4.12)

$$\bar{z}_3 = -A_{33}^{-1}A_{31}\bar{z}_1 - A_{33}^{-1}A_{32}\bar{z}_2 - A_{33}^{-1}B_3\bar{u} \tag{4.14}$$

where A_{33} is assumed to be nonsingular and $\bar{z}_1, \bar{z}_2, \bar{z}_3$ and $\bar{u}$ denote the values of the respective variables for the approximation $\mu = 0$.

Again, let $\varepsilon = 0$. Note that $\mu = 0$ when $\varepsilon = 0$ since $\mu \ll \varepsilon$. Then we have from (4.11) and (4.12)

$$A_{21}\tilde{z}_1 + A_{22}\tilde{z}_2 + A_{23}\tilde{z}_3 + B_2\tilde{u} = 0; \tag{4.15}$$

$$A_{31}\tilde{z}_1 + A_{32}\tilde{z}_2 + A_{33}\tilde{z}_3 + B_3\tilde{u} = 0, \tag{4.16}$$

where $\tilde{z}_1$, $\tilde{z}_2$, $\tilde{z}_3$, and $\tilde{u}$, denote the variables corresponding to the approximation $\varepsilon = 0$.

Assuming that $\left(A_{22} - A_{23}A_{33}^{-1}A_{32}\right)$ is non singular, (4.15) and (4.16) can be solved to obtain

$$\begin{aligned} \tilde{z}_2 = &- \left(A_{22} - A_{23}A_{33}^{-1}A_{32}\right)^{-1} \left(A_{21} - A_{23}A_{33}^{-1}A_{31}\right)\tilde{z}_1 \\ &- \left(A_{22} - A_{23}A_{33}^{-1}A_{32}\right)^{-1} \left(B_2 - A_{23}A_{33}^{-1}B_3\right)\tilde{u}, \end{aligned} \tag{4.17}$$

$$\text{and } \tilde{z}_3 = -A_{33}^{-1}A_{31}\tilde{z}_1 - A_{33}^{-1}A_{32}\tilde{x}_2 - A_{33}^{-1}B_3\tilde{u}. \tag{4.18}$$

Substituting (4.17) and (4.18) in set of equations (4.10)–(4.13), we get

$$\begin{aligned} \dot{\tilde{z}}_1 = &\left[A_{11} - A_{13}A_{33}^{-1}A_{31} - \left(A_{12} - A_{13}A_{33}^{-1}A_{32}\right)\left(A_{22} - A_{23}A_{33}^{-1}A_{32}\right)^{-1}\left(A_{21} - A_{23}A_{33}^{-1}A_{31}\right)\right]\tilde{z}_1 \\ &- \left[\left(A_{12} - A_{13}A_{33}^{-1}A_{32}\right)\left(A_{22} - A_{23}A_{33}^{-1}A_{32}\right)^{-1}\left(B_2 - A_{23}A_{33}^{-1}B_3\right) \right. \\ &\left. - B_1 + A_{13}A_{33}^{-1}B_3\right]\tilde{u} \end{aligned} \tag{4.19}$$

$$\begin{aligned} \tilde{y} = &\left[M_1 - M_3A_{33}^{-1}A_{31} - \left(M_2 - M_3A_{33}^{-1}A_{32}\right)\left(A_{22} - A_{23}A_{33}^{-1}A_{32}\right)^{-1}\left(A_{21} - A_{23}A_{33}^{-1}A_{31}\right)\right]\tilde{z}_1 \\ &- \left[M_3A_{33}^{-1}B_3 + \left(M_2 - M_3A_{33}^{-1}A_{32}\right)\left(A_{22} - A_{23}A_{33}^{-1}A_{32}\right)^{-1}\left(B_2 - A_{23}A_{33}^{-1}B_3\right)\right]\tilde{u}. \end{aligned} \tag{4.20}$$

where $\tilde{z}_1(0) = z_{1_0}$. Now, based on the above derivations, the three–time–scale system described by (4.10)–(4.13) can be decomposed into three separate subsystems, as follows.

4.2.1 Slow Subsystem

Defining $z_s = \tilde{z}_1$ and $u_s = \tilde{u}$, the slow subsystem is obtained from (4.19) as

$$\dot{z}_s = A_s z_s + B_s u_s, \tag{4.21}$$

$$y_s = M_s z_s + D_s u_s; \tag{4.22}$$

where,

$$\begin{aligned} A_s = A_{11} &- \left(A_{12} - A_{13}A_{33}^{-1}A_{32}\right)\left(A_{22} - A_{23}A_{33}^{-1}A_{32}\right)^{-1}\left(A_{21} - A_{23}A_{33}^{-1}A_{31}\right) \\ &- A_{13}A_{33}^{-1}A_{31}, \end{aligned} \tag{4.23}$$

$$B_s = B_1 - \left(A_{12} - A_{13}A_{33}^{-1}A_{32}\right)\left(A_{22} - A_{23}A_{33}^{-1}A_{32}\right)^{-1}\left(B_2 - A_{23}A_{33}^{-1}B_3\right) \\ -A_{13}A_{33}^{-1}B_3, \tag{4.24}$$

$$M_s = M_1 - \left(M_2 - M_3A_{33}^{-1}A_{32}\right)\left(A_{22} - A_{23}A_{33}^{-1}A_{32}\right)^{-1}\left(A_{21} - A_{23}A_{33}^{-1}A_{31}\right) \\ -M_3A_{33}^{-1}A_{31}, \tag{4.25}$$

$$D_s = -\left(M_2 - M_3A_{33}^{-1}A_{32}\right)\left(A_{22} - A_{23}A_{33}^{-1}A_{32}\right)^{-1}\left(B_2 - A_{23}A_{33}^{-1}B_3\right) \\ -M_3A_{33}^{-1}B_3. \tag{4.26}$$

4.2.2 *Fast 1 Subsystem*

Similarly, by defining the ε-stretched time scale as $\tau_\varepsilon = \frac{(t-t_0)}{\varepsilon}$, in which $z_1 = z_{1_0}$, $z_3 = \bar{z}_3$ and by introducing $z_{f1} = z_2 - \tilde{z}_2$, $u_{f1} = \bar{u} - \tilde{u}$ and $y_{f1} = y - \tilde{y}$, we obtain,

$$\frac{dz_{f1}}{d\tau_\varepsilon} = A_{f1}z_{f1} + B_{f1}u_{f1}; \tag{4.27}$$

$$y_{f1} = M_{f1}z_{f1} + D_{f1}u_{f1}; \tag{4.28}$$

where $A_{f1} = A_{22} - A_{23}A_{33}^{-1}A_{32}$, $B_{f1} = B_2 - A_{23}A_{33}^{-1}B_3$, $M_{f1} = M_2 - M_3A_{33}^{-1}A_{32}$ and $D_{f1} = -M_3A_{33}^{-1}B_3$.

4.2.3 *Fast 2 Subsystem*

By adopting the μ-stretched time scale defined as $\tau_\mu = \frac{(t-t_0)}{\mu}$, in which the fast 1 and slow subsystem states are assumed to be constant, *i.e.*,

$$\frac{dz_1}{d\tau_\mu} = 0, \quad \frac{dz_2}{d\tau_\mu} = 0 \;\Rightarrow\; z_1 = z_{1_0},\; z_2 = z_{2_0},$$

we get

$$\frac{dz_3}{d\tau_\mu} = A_{31}z_{1_0} + A_{32}z_{2_0} + A_{33}z_3 + B_3u, \tag{4.29}$$

$$y = M_1z_{1_0} + M_2z_{2_0} + M_3z_3.$$

Let $z_{f2} = z_3 - \bar{z}_3$, $u_{f2} = u - \bar{u}$ and $y_{f2} = y - \bar{y}$, with which (5.1) becomes

$$\frac{dz_{f2}}{d\tau_\mu} = A_{f2}z_{f2} + B_{f2}u_{f2}, \tag{4.30}$$

$$y_{f2} = M_{f2}z_{f2}; \tag{4.31}$$

where $A_{f2} = A_{33}$, $B_{f2} = B_3$ and $M_{f2} = M_3$.

Thus the original high order system described by (4.10)–(4.13) is decomposed into a slow subsystem given by (4.21) and (4.22), a fast subsystem given by (4.27) and (4.28), and a very fast subsystem given by (4.30) and (4.31).

4.3 Design of Linear State Regulator

In this section, let us consider the design of linear state regulator for the system described by (4.10)–(4.13), rewritten for convenience as

$$\dot{z} = Az + Bu, \quad y = Mz, \tag{4.32}$$

where $z = [z_1^T \;\; z_2^T \;\; z_3^T]^T$ is the $n = n_1 + n_2 + n_3$ dimensional state vector, and

$$A = \begin{bmatrix} A_{11} & A_{12} & A_{13} \\ A_{21}/\varepsilon & A_{22}/\varepsilon & A_{23}/\varepsilon \\ A_{31}/\mu & A_{32}/\mu & A_{33}/\mu \end{bmatrix}, \tag{4.33}$$

$$B = \left[B_1^T \; B_2^T/\varepsilon \; B_3^T/\mu \right]^T, \tag{4.34}$$

$$\text{and } M = \left[M_1 \; M_2 \; M_3 \right]. \tag{4.35}$$

In particular, it is considered to minimize the cost function

$$J = \int_0^\infty \left(z^T Q z + u^T R u \right) dt \tag{4.36}$$

where $Q \geq 0$ and $R > 0$ are respectively $n \times n$ dimensional state weighting matrix and $m \times m$ input weighting matrix. The control law that minimizes J for (4.32) is given by

$$\hat{u} = -R^{-1} B^T P z = -K^* z, \tag{4.37}$$

where P is the positive definite solution of the algebraic matrix Riccati equation

$$A^T P + PA - PBR^{-1} B^T P + Q = 0. \tag{4.38}$$

The presence of time scales in the dynamics of the system (4.32) is known to cause ill–conditioning whereby it becomes difficult to determine P directly from (4.38). A two time–scale approach for design of linear state regulators for such systems has been suggested in [18].

In the following, a three–time–scale approach that utilizes the slow, fast 1 and fast 2 subsystem models derived in the preceeding section, is proposed for designing linear state regulators separately for these subsystems. Finally, the separately designed regulators are combined to obtain the control law (4.37). This approach obviates the ill–conditioning, and also three smaller order problems are to be solved instead of a high order problem. At the outset the matrices Q and P are assumed to be in the following form:

$$Q = \begin{bmatrix} Q_{11} & Q_{12} & Q_{13} \\ Q_{12}^T & Q_{22} & Q_{23} \\ Q_{13}^T & Q_{23}^T & Q_{33} \end{bmatrix}, \tag{4.39}$$

$$\text{and } P = \begin{bmatrix} P_{11} & \varepsilon P_{12} & \mu P_{13} \\ \varepsilon P_{12}^T & \varepsilon P_{22} & \mu P_{23} \\ \mu P_{13}^T & \mu P_{23}^T & \mu P_{33} \end{bmatrix}. \tag{4.40}$$

With this, three quadratic performance indices can be extracted from J, one each for slow, fast 1 and fast 2 subsystems respectively, as follows.

4.3.1 Slow Subsystem Regulator

With $\varepsilon = 0$ in (4.12), replacing $\tilde{z}_2$ and $\tilde{z}_3$ in terms of $z_s = \tilde{z}_1$ and $u_s = \tilde{u}$, the cost function for the slow subsystem can be obtained as

$$J_s = \int_0^\infty \left(z_s^T Q_s z_s + u_s^T R_s u_s + 2 z_s^T \tilde{Q}_s u_s\right) dt \tag{4.41}$$

where

$$\begin{aligned} Q_s &= Q_{11} + (Q_{12}\tilde{A}_{21})^T + (Q_{13}\tilde{A}_{31})^T + \left[Q_{12} + \tilde{A}_{21}^T Q_{22} + (Q_{23}\tilde{A}_{31})^T\right]\tilde{A}_{21} \\ &\quad + \left[Q_{13} + \tilde{A}_{21}^T Q_{23} + \tilde{A}_{31}^T Q_{33}\right]\tilde{A}_{31}, & (4.42) \\ R_s &= R + \left[\tilde{B}_2^T Q_{22} + (Q_{23}\tilde{B}_3)^T\right]\tilde{B}_2 + \left(\tilde{B}_2^T Q_{23} + \tilde{B}_3^T Q_{33}\right)\tilde{B}_3, & (4.43) \\ \tilde{Q}_s &= \left[Q_{12} + \tilde{A}_{21}^T Q_{22} + (Q_{23}\tilde{A}_{31})^T\right]\tilde{B}_2 + \left(Q_{13} + \tilde{A}_{21}^T Q_{23} + \tilde{A}_{31}^T Q_{33}\right)\tilde{B}_3, & (4.44) \end{aligned}$$

in which

$$\begin{aligned} \tilde{A}_{21} &= -\left(A_{22} - A_{23}A_{33}^{-1}A_{32}\right)^{-1}\left(A_{21} - A_{23}A_{33}^{-1}A_{31}\right), & (4.45) \\ \tilde{A}_{31} &= -A_{33}^{-1}\left(A_{32}\tilde{A}_{21} + A_{31}\right), & (4.46) \\ \tilde{B}_2 &= -\left(A_{22} - A_{23}A_{33}^{-1}A_{32}\right)^{-1}\left(B_2 - A_{23}A_{33}^{-1}B_3\right), & (4.47) \\ \tilde{B}_3 &= -A_{33}^{-1}\left(B_3 + A_{32}\tilde{B}_2\right). & (4.48) \end{aligned}$$

The solution of the problem is

$$u_s = -R_s^{-1}\left(P_s B_s + \tilde{Q}_s\right)^T z_s = -K_s z_s, \tag{4.49}$$

where $P_s = P_{11}$ satisfies the Riccati equation

$$A_s^T P_s + P_s A_s - N_s R_s^{-1} N_s^T + Q_s = 0, \tag{4.50}$$

in which $N_s = P_s B_s + \tilde{Q}_s$.

4.3.2 Fast 1 Subsystem Regulator

Find u_{f1} to minimize

$$J_{f1} = \int_0^\infty \left(z_{f1}^T Q_{f1} z_{f1} + u_{f1}^T R_{f1} u_{f1} + 2 z_{f1}^T \tilde{Q}_{f1} u_{f1} \right) dt$$

for the fast 1 subsystem (4.27), where

$$\tilde{Q}_{f1} = \left(A_{33}^{-1} A_{32}\right)^T Q_{33} \left(A_{33}^{-1} B_3\right) - Q_{23} \left(A_{33}^{-1} B_3\right), \tag{4.51}$$

$$Q_{f1} = Q_{22} - Q_{23} A_{33}^{-1} A_{32} - \left(Q_{23} A_{33}^{-1} A_{32}\right)^T - \left(A_{33}^{-1} A_{32}\right)^T Q_{33} \left(A_{33}^{-1} A_{32}\right), \tag{4.52}$$

$$R_{f1} = R + \left(A_{33}^{-1} B_3\right)^T Q_{33} \left(A_{33}^{-1} B_3\right). \tag{4.53}$$

The optimal solution for this problem is

$$u_{f1} = -R_{f1}^{-1} \left(P_{f1} B_{f1} + \tilde{Q}_{f1}\right)^T z_{f1} = -K_{f1} z_{f1}, \tag{4.54}$$

in which $P_{f1} = P_{22}$ is obtained from the Riccati equation

$$A_{f1}^T P_{f1} + P_{f1} A_{f1} - N_{f1} R_{f1}^{-1} N_{f1}^T + Q_{f1} = 0, \tag{4.55}$$

where

$$N_{f1} = P_{f1} \left(B_2 - A_{23} A_{33}^{-1} B_3\right) + \tilde{Q}_{f1}. \tag{4.56}$$

4.3.3 Fast 2 Subsystem Regulator

For the fast 2 subsystem (4.30), the optimal control problem is obtained as finding u_{f2} to minimize

$$J_{f2} = \int_0^\infty \left(z_{f2}^T Q_{f2} z_{f2} + u_{f2}^T R u_{f2} \right) dt$$

where, in terms of the Q matrix for the original system, $Q_{f2} = Q_{33}$, and the optimal control for the problem is

$$u_{f2} = -R^{-1} B_{f2}^T P_{f2} z_{f2} = -K_{f2} z_{f2}, \tag{4.57}$$

where $P_{f2} = P_{33}$ is the solution of the Riccati equation

$$A_{f2}^T P_{f2} + P_{f2} A_{f2} - P_{f2} B_{f2} R^{-1} B_{f2}^T P_{f2} + Q_{f2} = 0. \tag{4.58}$$

The derivation of slow, fast 1 and fast 2 subsystem regulators is given in Appendix.

4.3.4 *Composite Control*

Assume that the fast 2 subsystem reaches its quasi-steady state very fast and remains constant afterwards, such that $\bar{z}_3 = \tilde{z}_3$. Also recall that $z_{f2} = z_3 - \bar{z}_3$, $z_{f1} = z_2 - \tilde{z}_2$ and $z_s = \tilde{z}_1 = z_1$. Then the composite control for the original system is $u = u_{f1} + u_{f2} + u_s$, or,

$$\begin{aligned} u = {} & [K_s + K_{f2} A_{33}^{-1} (A_{31} + B_3 K_s) \\ & + (K_{f1} - K_{f2} A_{33}^{-1} A_{32}) (A_{22} - A_{23} A_{33}^{-1} A_{32})^{-1} (A_{21} - A_{23} A_{33}^{-1} (A_{31} + B_3 K_s) + B_2 K_s)] z_1 \\ & + K_{f1} z_2 + K_{f2} z_3 \quad = -Kz, \end{aligned} \tag{4.59}$$

which is readily derived using (4.17), (4.18), (4.49), (4.54) and (4.57).

It may be noticed that on discarding the fast 1 subsystem, (4.59) reduces to the composite controller for two–time scale singularly perturbed systems as derived in [18].

4.4 Application to the AHWR Model

As mentioned in the previous chapter, the open loop eigenvalues of the AHWR model fall in three distinct clusters. Even with the total power feedback, this property is preserved. This three–time–scale property of the model could be exploited to decompose the large order system into three smaller order subsystems so as to eliminate the numerical ill–conditioning effects. Hence, in lines with the steps described in Sec. 4.2, the AHWR model with total power feedback, described by (3.3), is decomposed into a slow subsystem of order 38, a fast 1 subsystem of order 35 and a fast 2 subsystem of order 17 [95]. Slow subsystem contains the eigenvalues which are unstable, alongwith those near the origin, whereas fast 1 and fast 2 subsystems contains only stable eigenvalues, as depicted in Table 4.1. It can be observed that the original system eigenvalues are the union of the subsystem eigenvalues. However, it turns out that the fast 1 and fast 2 subsystems are fully uncontrollable since their input matrices are null, essentially due to null matrices B_3 and B_2, and the slow subsystem is controllable. Hence the application turns out to be a reduced control, with only K_s being designed with which the control law (4.59) becomes

$$u = K_s z. \tag{4.60}$$

Table 4.1 Comparison of Original and Subsystem Eigenvalues of the AHWR Model.

i	φ_i(orig. system)	φ_i(subsystem)	i	φ_i(orig. system)	φ_i(subsystem)
		Slow Subsystem			
$1-2$	$(8.8268 \pm j1.8656)\times 10^{-5}$	$(8.8423e-005+1.9101)\times 10^{-5}$	51	-6.2200×10^{-2}	-6.2198×10^{-2}
$3-4$	$(8.0470 \pm j2.4129)\times 10^{-5}$	$(8.0601e-005+2.4767)\times 10^{-5}$	52	-6.2380×10^{-2}	-6.2379×10^{-2}
5	-1.8988×10^{-8}	-1.8010×10^{-8}	53	-6.2458×10^{-2}	-6.2456×10^{-2}
6	-2.0479×10^{-8}	-1.9120×10^{-8}	54	-6.2608×10^{-2}	-6.2607×10^{-2}
7	-6.9610×10^{-8}	-6.3513×10^{-8}	55	-6.2865×10^{-2}	-6.2864×10^{-2}
$8-9$	$(-3.6138 \pm j7.6617)\times 10^{-5}$	$(-3.6150e-005+7.6626)\times 10^{-5}$	56	-6.2893×10^{-2}	-6.2893×10^{-2}
10	-3.7781×10^{-5}	-3.7780×10^{-5}	57	-1.1714×10^{-1}	-1.1714×10^{-1}
$11-12$	$(-3.7785 \pm j7.6475)\times 10^{-5}$	$(-3.7798e-005+7.6484)\times 10^{-5}$	58	-1.4712×10^{-1}	-1.4712×10^{-1}
13	-3.7993×10^{-5}	-3.7993×10^{-5}	59	-1.4713×10^{-1}	-1.4713×10^{-1}
14	-4.0124×10^{-5}	-4.0124×10^{-5}	60	-1.4809×10^{-1}	-1.4808×10^{-1}
15	-4.1520×10^{-5}	-4.1520×10^{-5}	61	-1.4850×10^{-1}	-1.4849×10^{-1}
16	-4.2219×10^{-5}	-4.2218×10^{-5}	62	-1.5580×10^{-1}	-1.5580×10^{-1}
17	-4.4204×10^{-5}	-4.4203×10^{-5}	63	-1.5585×10^{-1}	-1.5585×10^{-1}
18	-4.6941×10^{-5}	-4.6940×10^{-5}	64	-1.5662×10^{-1}	-1.5662×10^{-1}
19	-4.8866×10^{-5}	-4.8865×10^{-5}	65	-1.5760×10^{-1}	-1.5758×10^{-1}
$20-21$	$(-6.4855 \pm j5.3109)\times 10^{-5}$	$(-6.4866e-005+5.3107)\times 10^{-5}$	66	-1.6316×10^{-1}	-1.6316×10^{-1}
$22-23$	$(-6.6069 \pm j5.4962)\times 10^{-5}$	$(-6.6081e-005+5.4960)\times 10^{-5}$	67	-1.6325×10^{-1}	-1.6324×10^{-1}
$24-25$	$(-7.3359 \pm j3.9319)\times 10^{-5}$	$(-7.3368e-005+3.9313)\times 10^{-5}$	68	-1.6405×10^{-1}	-1.6405×10^{-1}
$26-27$	$(-7.7407 \pm j2.9929)\times 10^{-5}$	$(-7.7415e-005+2.9920)\times 10^{-5}$	69	-1.6578×10^{-1}	-1.6573×10^{-1}
28	-1.4107×10^{-4}	-1.4108×10^{-4}	70	-1.8037×10^{-1}	-1.8037×10^{-1}
29	-1.4601×10^{-4}	-1.4602×10^{-4}	71	-1.8049×10^{-1}	-1.8049×10^{-1}
30	-1.5717×10^{-4}	-1.5718×10^{-4}	72	-1.8122×10^{-1}	-1.8121×10^{-1}
31	-1.6524×10^{-4}	-1.6524×10^{-4}	73	-1.8399×10^{-1}	-1.8390×10^{-1}
32	-1.6648×10^{-4}	-1.6649×10^{-4}			Fast 2 Subsystem
33	-1.7308×10^{-4}	-1.7309×10^{-4}	74	-7.2482	-7.2028
34	-1.8807×10^{-4}	-1.8807×10^{-4}	75	-3.2844×10^{1}	-3.2833×10^{1}
35	-1.8870×10^{-4}	-1.8871×10^{-4}	76	-3.3372×10^{1}	-3.3361×10^{1}
36	-2.6916×10^{-4}	-2.6919×10^{-4}	77	-6.6599×10^{1}	-6.6593×10^{1}
37	-4.0318×10^{-4}	-4.0315×10^{-4}	78	-6.8323×10^{1}	-6.8317×10^{1}
38	-4.8727×10^{-4}	-4.8721×10^{-4}	79	-9.3653×10^{1}	-9.3649×10^{1}
		Fast 1 Subsystem	80	-9.4612×10^{1}	-9.4608×10^{1}
39	-5.9378×10^{-2}	-6.5475×10^{-2}	81	-1.0868×10^{2}	-1.0868×10^{2}
40	-1.5924×10^{-2}	-1.5933×10^{-2}	82	-1.1705×10^{2}	-1.1704×10^{2}
41	-5.0954×10^{-2}	-5.0882×10^{-2}	83	-1.6967×10^{2}	-1.6967×10^{2}
42	-5.1159×10^{-2}	-5.1093×10^{-2}	84	-1.7568×10^{2}	-1.7568×10^{2}
43	-5.7731×10^{-2}	-5.7716×10^{-2}	85	-1.9497×10^{2}	-1.9497×10^{2}
44	-5.7893×10^{-2}	-5.7879×10^{-2}	86	-2.1110×10^{2}	-2.1110×10^{2}
45	-5.9707×10^{-2}	-5.9700×10^{-2}	87	-2.1904×10^{2}	-2.1904×10^{2}
46	-5.9723×10^{-2}	-5.9717×10^{-2}	88	-2.3591×10^{2}	-2.3591×10^{2}
47	-6.0344×10^{-2}	-6.0339×10^{-2}	89	-2.7163×10^{2}	-2.7163×10^{2}
48	-6.0642×10^{-2}	-6.0638×10^{-2}	90	-2.7626×10^{2}	-2.7626×10^{2}
49	-6.1848×10^{-2}	-6.1846×10^{-2}			
50	-6.1942×10^{-2}	-6.1940×10^{-2}			

Q and R matrices for the original system were chosen as $10\times E_{90}$ and $1000\times E_4$ respectively, where E_i denotes an identity matrix of dimension i. With this, subsystem regulator problem was formulated for the slow subsystem. A composite regulator of dimension 4×90 is obtained for the original system from (4.60). The gain matrix thus obtained has non–zero entries only in the first 38 columns, with magnitudes varying from the order of 10^{-3} to 10^{0}. Table 4.2 lists the closed loop eigenvalues of the AHWR with this reduced, stabilizing, composite regulator.

Table 4.2 Closed Loop Eigenvalues of the AHWR Model.

i	φ_i	i	φ_i
1	-2.8227×10^{-5}	49	-6.1848×10^{-2}
2	-2.8757×10^{-5}	50	-6.1943×10^{-2}
3	-2.8772×10^{-5}	51	-6.2200×10^{-2}
4	-2.8773×10^{-5}	52	-6.2380×10^{-2}
5	-3.5825×10^{-5}	53	-6.2458×10^{-2}
$6-7$	$(-3.7554 \pm j7.6516) \times 10^{-5}$	54	-6.2608×10^{-2}
8	-3.8272×10^{-5}	55	-6.2865×10^{-2}
9	-4.0750×10^{-5}	56	-6.2893×10^{-2}
10	-4.0813×10^{-5}	57	-1.1714×10^{-1}
11	-4.2983×10^{-5}	58	-1.4712×10^{-1}
12	-4.3190×10^{-5}	59	-1.4713×10^{-1}
13	-4.5833×10^{-5}	60	-1.4809×10^{-1}
14	-4.7219×10^{-5}	61	-1.4850×10^{-1}
$15-16$	$(-6.5904 \pm j5.5060) \times 10^{-5}$	62	-1.5580×10^{-1}
$17-18$	$(-6.6211 \pm j5.1794) \times 10^{-5}$	63	-1.5585×10^{-1}
$19-20$	$(-7.4102 \pm j3.8304) \times 10^{-5}$	64	-1.5662×10^{-1}
$21-22$	$(-8.1483 \pm j1.5267) \times 10^{-5}$	65	-1.5760×10^{-1}
23	-1.4343×10^{-4}	66	-1.6316×10^{-1}
24	-1.4678×10^{-4}	67	-1.6325×10^{-1}
25	-1.5511×10^{-4}	68	-1.6405×10^{-1}
26	-1.5694×10^{-4}	69	-1.6578×10^{-1}
27	-1.6856×10^{-4}	70	-1.8037×10^{-1}
28	-1.6950×10^{-4}	71	-1.8049×10^{-1}
29	-1.8710×10^{-4}	72	-1.8122×10^{-1}
30	-1.9567×10^{-4}	73	-1.8399×10^{-1}
31	-2.3825×10^{-4}	74	-7.2499
32	-2.5231×10^{-4}	75	-3.2844×10^{1}
33	-2.5258×10^{-4}	76	-3.3372×10^{1}
34	-2.6869×10^{-4}	77	-6.6599×10^{1}
35	-7.4343×10^{-4}	78	-6.8323×10^{1}
36	-1.3733×10^{-3}	79	-9.3653×10^{1}
37	-1.4989×10^{-3}	80	-9.4612×10^{1}
38	-6.6281×10^{-3}	81	-1.0868×10^{2}
39	-1.5934×10^{-2}	82	-1.1705×10^{2}
40	-4.7350×10^{-2}	83	-1.6967×10^{2}
41	-5.0958×10^{-2}	84	-1.7568×10^{2}
42	-5.1162×10^{-2}	85	-1.9497×10^{2}
43	-5.7733×10^{-2}	86	-2.1110×10^{2}
44	-5.7893×10^{-2}	87	-2.1904×10^{2}
45	-5.9707×10^{-2}	88	-2.3591×10^{2}
46	-5.9723×10^{-2}	89	-2.7163×10^{2}
47	-6.0344×10^{-2}	90	-2.7626×10^{2}
48	-6.0642×10^{-2}		

4.4.1 *Transient Response*

Response of the controller was analyzed by simulating a transient involving a disturbance in the spatial power distribution. The system was assumed to be at full power steady state operation prior to the transient. Shortly afterwards, RR8 was manually driven out by about 1.5% under a control signal of $-1V$. Parallely RR4 was manually driven in under a signal of $1V$. The RRs were left at their new positions to the effect of the spatial power controller. It was observed that the RRs were driven back to their equilibrium positions by the controller, as shown in Fig. 4.1. As a result of the controller action, disturbances in the nodal powers as well as the quadrant core power distribution were suppressed within about 200 s, as depicted in Fig. 4.2 and 4.3 respectively.

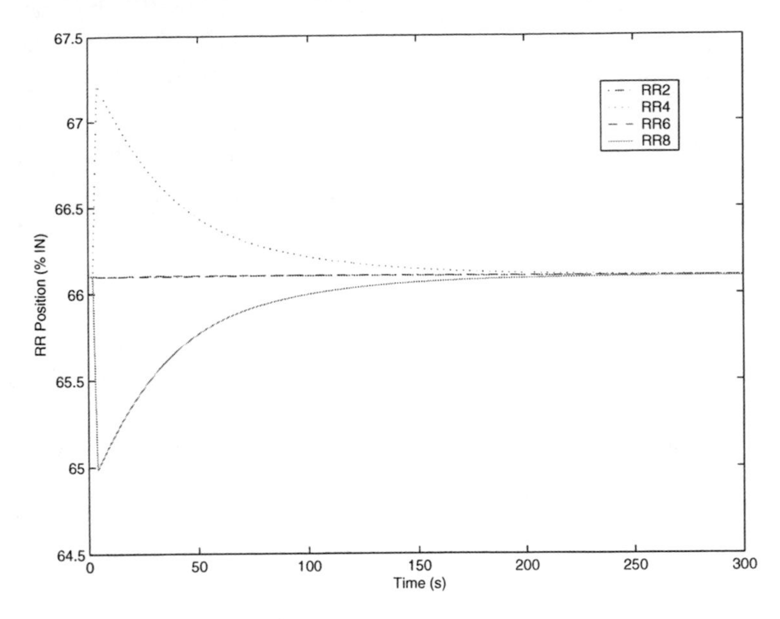

Fig. 4.1 RR Positions.

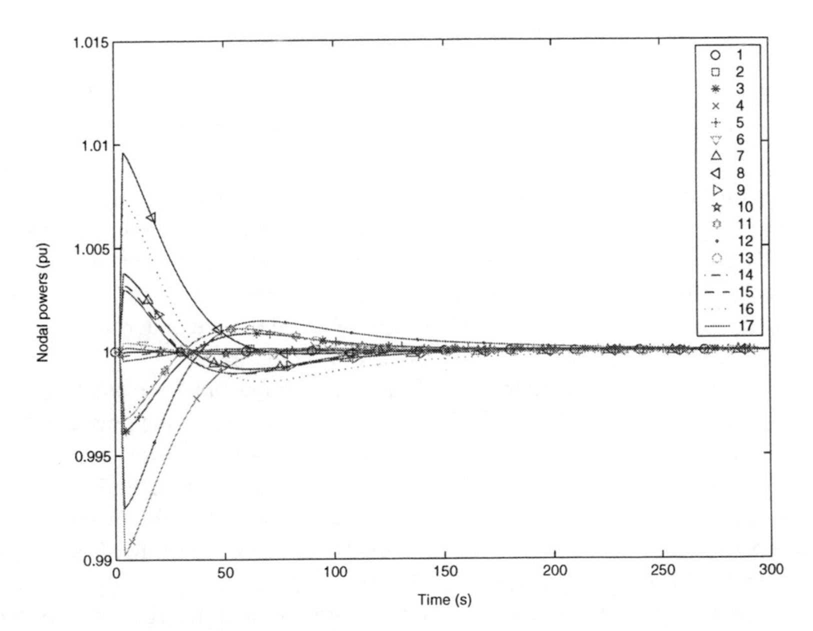

Fig. 4.2 Nodal Power Variations.

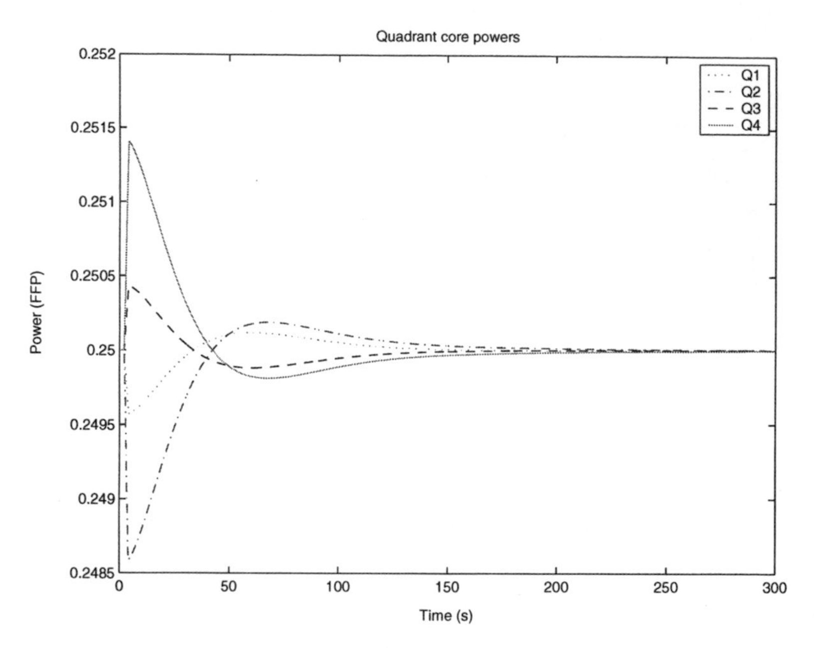

Fig. 4.3 Quadrant Core Power Variations.

Transient response of the closed loop system with the control law (4.60) was assessed for the transient mentioned in Sec. 3.4 as well. The dynamic equations were solved with a time step of 5 ms. The reactor was assumed to be initially operating at full power equilibrium conditions. Shortly, RR2, originally under auto control, was manually driven out by about 2% under a signal of $-1V$, and left on auto control afterwards. RR4, RR6 and RR8 were under auto control and total and spatial power controllers were active throughout. Other four RRs which are under manual control are assumed to remain stationary throughout the simulation.

During the manual retraction of RR2, it was observed that the total power experienced an increase by about 0.32%. However, subsequently the power stabilized back at the rated value, as shown in Fig. 4.4. Positions of all RRs are depicted in Fig. 4.5. It may be noticed that while RR2 is being driven out manually, other RRs moves in under the effect of the controller in order to maintain the total power. After the period during which the manual signal was enforced on RR2, all the RRs are being driven back by the controller to their original positions. Fig. 4.7 depicts the variations in normalized quadrant core powers, from which it may be observed that the power in the first quadrant which contains RR2 experienced an increase by about 0.1%, and it is being compensated by decrement in power generated in other quadrants of the core. The disturbances in the quadrant core powers are suppressed by the controller within about 250 seconds. It can be observed from Fig. 4.6 that the perturbations in nodal powers are also being eliminated by the controller. Furthermore, on continuing the simulation over an extended period, it was observed that the

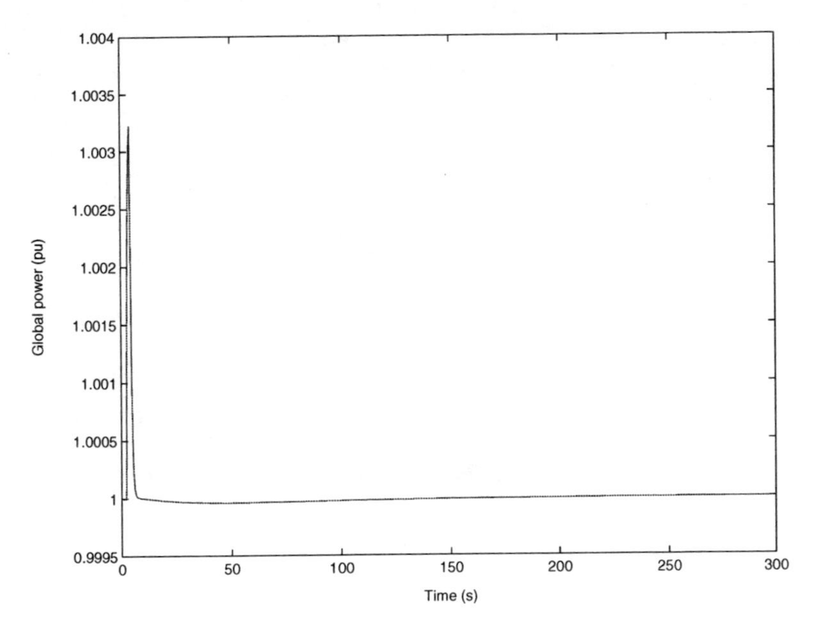

Fig. 4.4 Total Power Variation.

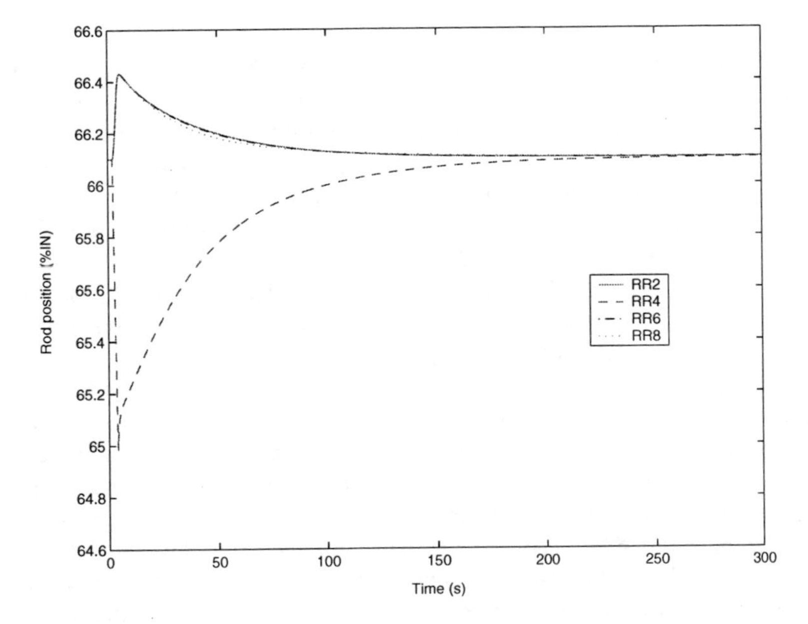

Fig. 4.5 RR Positions.

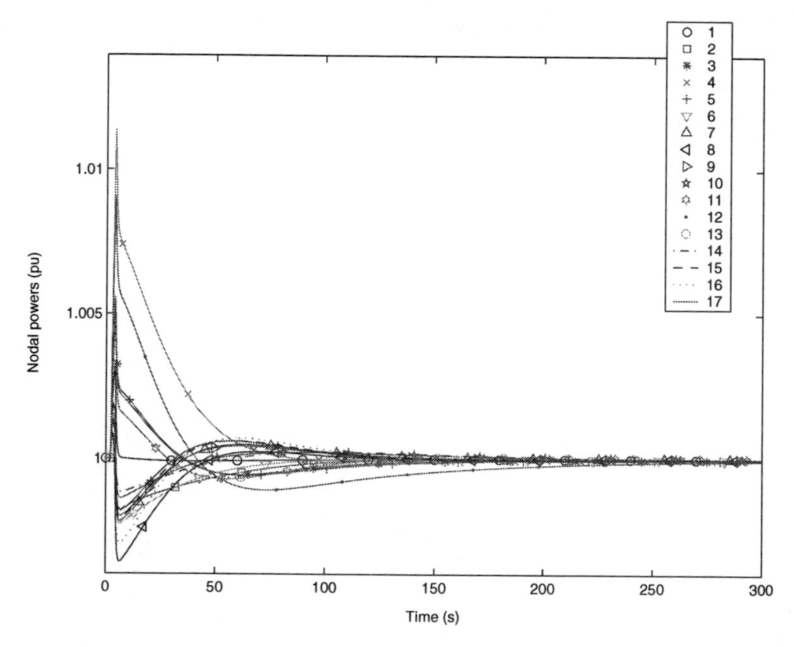

Fig. 4.6 Nodal Power Variations.

power tilts are not getting surfaced. However, it could be noticed that the response of the controller is slow as compared to that of the output feedback controller derived in Chapter 4, essentially since the former is a reduced order controller.

Fig. 4.8 shows the closed loop system response during another spatial power transient initiated by a momentary disturbance in positions of $RR4$ and $RR8$. The simulation was carried out with the non–linear model with a time step of 5 ms and data logging interval of 5 minutes. It was assumed that the reactor was initially operating at full power, with the control signal generated according to (3.1). The $RR8$, which was initially at its equilibrium position, was driven out by about 1.5% under a control signal of $-1V$, and $RR4$ was parallely driven in by the same amount under a signal of $1V$. Immediately afterwards, these RRs were driven back to their original positions under signals of $1V$ and $-1V$ respectively and thereafter again transferred under the influence of the controller. As a consequence of this disturbance and as the control is based only on the feedback of total power, the tilts started picking up. After about 260 minutes, the spatial control component was introduced in the control signals. Fig. 4.8(a) shows the variations in first azimuthal, second azimuthal and first radial tilts and Fig. 4.8(b) shows a zoomed version of the former, focusing the region near the introduction of spatial control. It was observed that the tilts are suppressed within about 10 minutes.

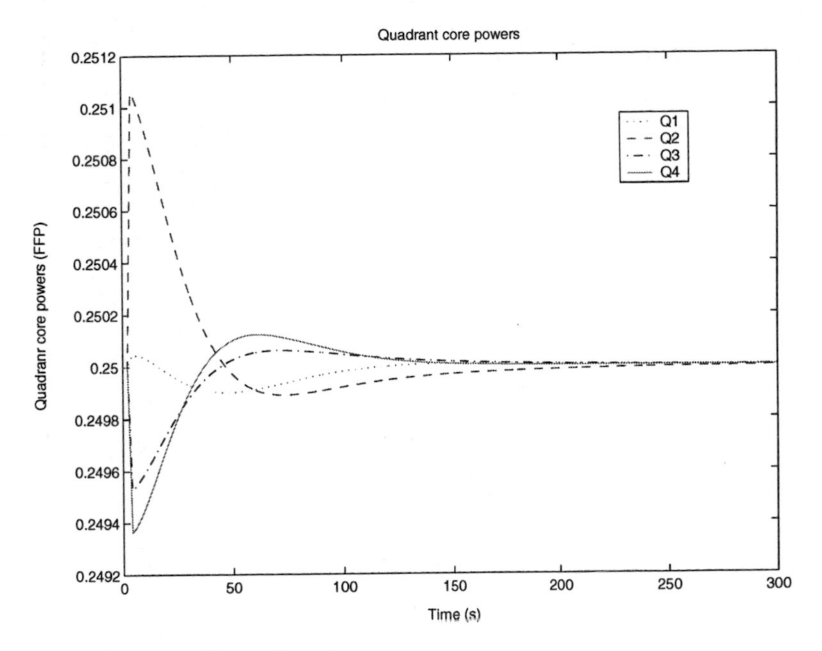

Fig. 4.7 Quadrant Core Power Variations.

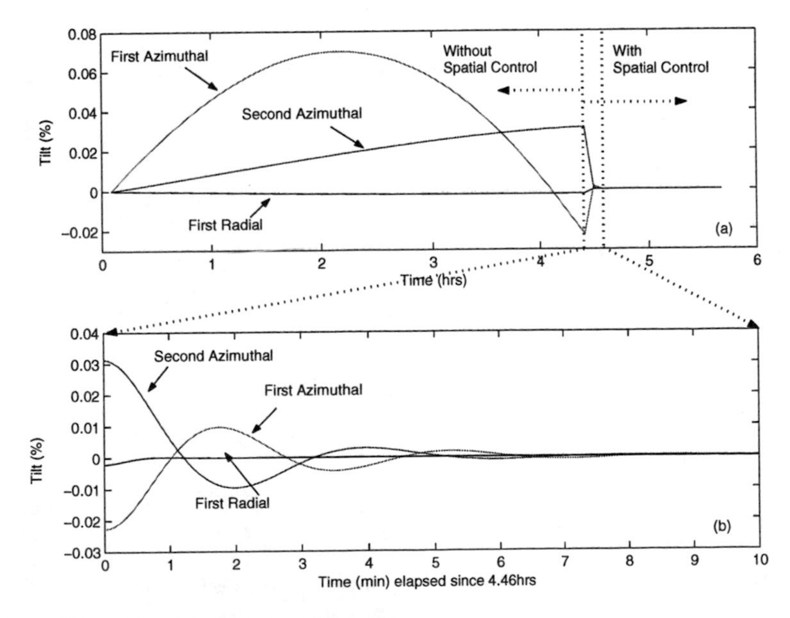

Fig. 4.8 Effect of Spatial Control on Flux Tilts.

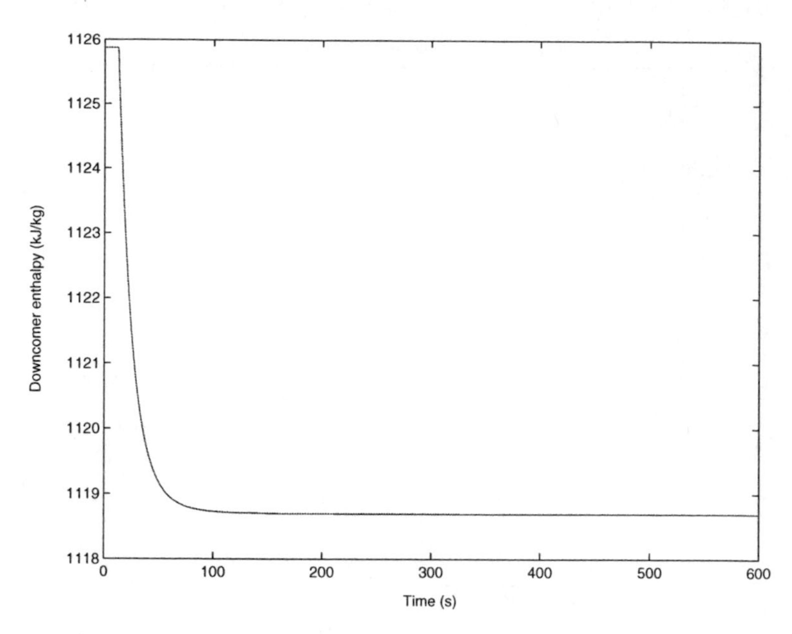

Fig. 4.9 Change in Downcomer Enthalpy.

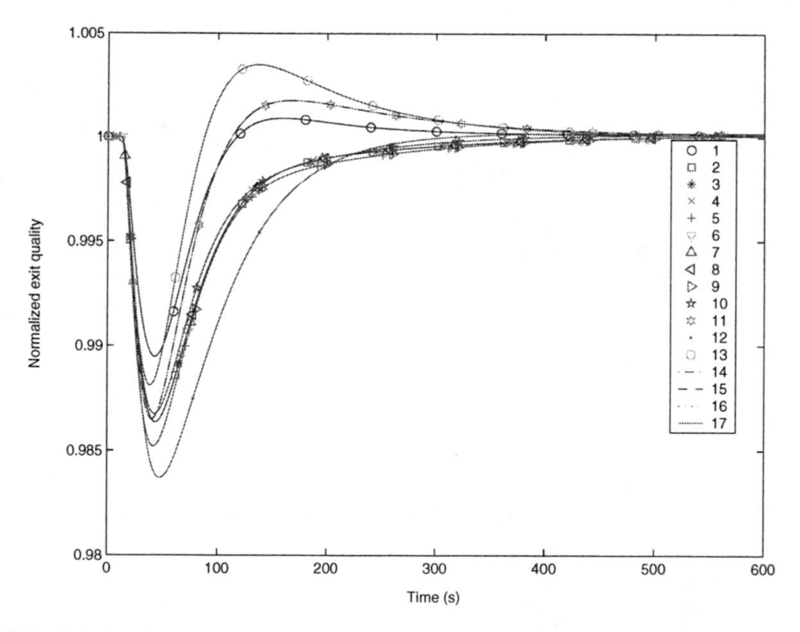

Fig. 4.10 Exit Quality Variations due to Disturbance in Feed Flow.

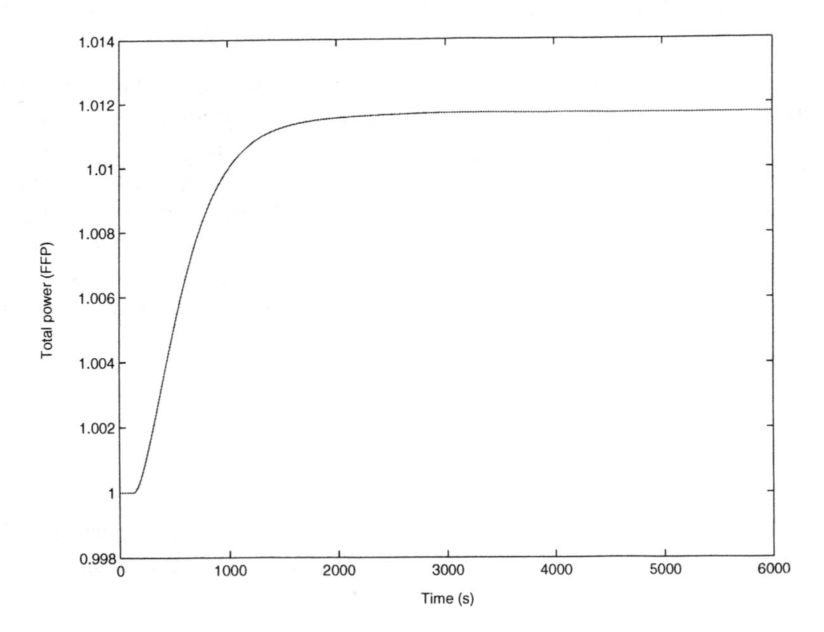

Fig. 4.11 Total Power Variation.

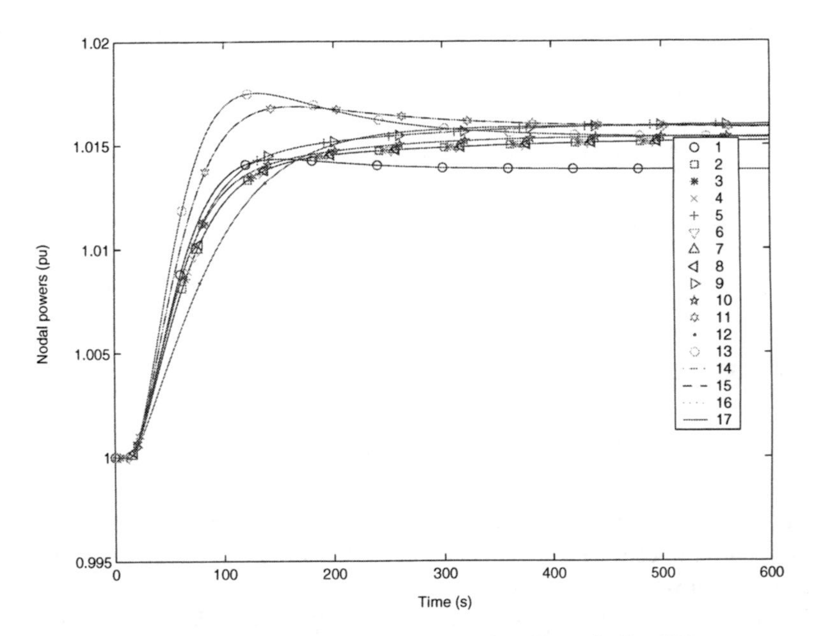

Fig. 4.12 Nodal Power Variations subsequent to the Disturbance in Feed Flow.

4.4.2 Robustness of the Controller

In order to assess the response of the system to disturbances in feed flow, a scenario identical to that described in Sec. 3.4.1 was simulated in which the reactor was operating at steady full power when a 5% positive step change was introduced in the feed flow. As a result of this, the incoming coolant enthalpy reduced by about 0.6% as shown in Fig. 4.9 and exit qualities from each node underwent changes as shown in Fig. 4.10. Total as well as nodal powers were found to be stabilizing at power levels of about 1.5% higher than the respective original values, as shown in Fig. 4.11 and Fig. 4.12. However, the transients were observed to be less oscillatory as compared to those reported in Sec. 3.4.1.

4.5 Conclusion

In short, the method described here achieves a direct single–stage decomposition of a large order system into three subsystems, unlike the methods involving two–stage application of two–time–scale decomposition technique. It is further demonstrated that the problem of linear regulator design for the original system can be decomposed into three smaller order problems, one each for individual subsystems, with independent quadratic performance indices. The proposed composite controller for three–time–scale systems, in terms of the three subsystem controllers, is a novel idea. However, analytic proof of the near optimality of the composite controller is yet to be established.

The proposed approach of three–time–scale decomposition and composite control can eliminate numerical ill–conditioning issues common with systems of large order and multiple time scales. The technique is applied to AHWR, where multi–time–scale space time kinetics model of very large order is involved, to obtain a stabilizing regulator.

Appendix

Derivation of Subsystem Regulators

The derivation of the subsystem optimum regulators involves simple but very lengthy calculations in lines with that explained in [35]. A brief discussion of the derivation procedure follows.

The substitution of (4.33), (4.34) and (4.40) in (4.38) yields at $\varepsilon = 0$ and $\mu = 0$ the equations

$$0 = A_{11}^T P_{11} + A_{21}^T P_{12}^T + A_{31}^T P_{13}^T + P_{11}A_{11} + P_{12}A_{21} + P_{13}A_{31} - NR^{-1}N^T + Q_{11} \quad (4.61)$$
$$0 = A_{22}^T P_{22} + A_{32}^T P_{23}^T + P_{22}A_{22} + P_{23}A_{32} - OR^{-1}O^T + Q_{22} \quad (4.62)$$
$$0 = A_{33}^T P_{33} + P_{33}A_{33} - P_{33}B_3R^{-1}B_3^T P_{33} + Q_{33} \quad (4.63)$$
$$0 = A_{21}^T P_{22} + A_{31}^T P_{23}^T + P_{11}A_{12} + P_{12}A_{22} + P_{13}A_{32} - NR^{-1}O^T + Q_{12} \quad (4.64)$$
$$0 = A_{13}^T P_{11} + A_{23}^T P_{12}^T + A_{33}^T P_{13}^T + P_{33}A_{31} - P_{33}B_3R^{-1}N^T + Q_{13}^T \quad (4.65)$$
$$0 = A_{32}^T P_{33} + P_{22}A_{23} + P_{23}A_{33} - OR^{-1}B_3^T P_{33} + Q_{23} \quad (4.66)$$

where $N = (P_{11}B_1 + P_{12}B_2 + P_{13}B_3)$ and $O = (P_{22}B_2 + P_{23}B_3)$.

Comparing (4.63) with (4.58) yields $P_{f2} = P_{33}$ and $Q_{f2} = Q_{33}$.

(4.66) gives

$$P_{23} = \left(-A_{32}^T P_{33} - P_{22}A_{23} + OR^{-1}B_3^T P_{33} - Q_{23}\right) A_{33}^{-1}. \quad (4.67)$$

Substituting (4.67) in (4.62) and eliminating the terms containing P_{33} by using (4.63) to obtain

$$A_{f1}^T P_{22} + P_{22}A_{f1} + Q_{22} - Q_{23}A_{33}^{-1}A_{32} - \left(Q_{23}A_{33}^{-1}A_{32}\right)^T - \left(A_{33}^{-1}A_{32}\right)^T Q_{33}\left(A_{33}^{-1}A_{32}\right) - HR^{-1}H^T = 0, \quad (4.68)$$

$$\text{where } H = O - \left(A_{33}^{-1}A_{32}\right) P_{33}B_3. \quad (4.69)$$

Substituting O and (4.67) in (4.69) results in

$$H = N_{f1}O_{f1} \quad \text{where} \quad (4.70)$$
$$N_{f1} = P_{22}\left(B_2 - A_{23}A_{33}^{-1}B_3\right) + \left(A_{33}^{-1}A_{32}\right)^T Q_{33}\left(A_{33}^{-1}B_3\right) - Q_{23}\left(A_{33}^{-1}B_3\right), \quad (4.71)$$
$$O_{f1} = \left(I - R^{-1}B_3^T P_{33}A_{33}^{-1}B_3\right)^{-1}. \quad (4.72)$$

Expression for O_{f1} can further be simplified with the use of (4.63) as

$$\begin{aligned} O_{f1}^{-1} &= I - R^{-1}B_3^T P_{33}A_{33}^{-1}B_3 \\ &= I - R^{-1}B_3^T A_{33}^T\left(P_{33}B_3R^{-1}B_3^T P_{33} - Q_{33} - P_{33}A_{33}\right)A_{33}^{-1}B_3 \\ &= R^{-1}R_{f1} + R^{-1}\tilde{C}_{f1}^T P_{33}B_3O_{f1}^{-1} \end{aligned}$$

where $\tilde{C}_{f1} = A_{33}^{-1}B_3$ and $R_{f1} = R + \tilde{C}_{f1}^T Q_{33}\tilde{C}_{f1}$. This implies that

$$O_{f1} = R_{f1}^{-1}R\left(I - R^{-1}\tilde{C}_{f1}^T P_{33}B_3\right), \text{ or,}$$
$$O_{f1}R^{-1}O_{f1}^T = R_{f1}^{-1},$$

which yields $HR^{-1}H^T = N_{f1}R_{f1}^{-1}N_{f1}^T$, with which (4.68) becomes

$$A_{f1}^T P_{22} + P_{22}A_{f1} + Q_{f1} - N_{f1}R_{f1}^{-1}N_{f1}^T = 0. \quad (4.73)$$

Comparison of (4.73) with (4.55) shows that $P_{f1} = P_{22}$.

From (4.65) it is obtained

$$P_{13} = \left(-P_{11}A_{13} - P_{12}A_{23} - A_{31}^T P_{33} + NR^{-1}B_3^T P_{33} - Q_{13}\right) A_{33}^{-1}. \tag{4.74}$$

Substituting (4.74) and (4.67) in (4.64) yields

$$\begin{aligned} P_{12} = &\Big[\left(A_{33}^{-1}A_{31}\right)^T P_{33}B_3R^{-1}\left(B_3^T P_{33} - O^T\right) - NR^{-1}B_3^T P_{33}\left(A_{33}^{-1}A_{32}\right) + NR^{-1}O^T \\ &+ \left(Q_{23}A_{33}^{-1}A_{31}\right)^T + Q_3 A_{33}^{-1}A_{32} - Q_{12} - \left(A_{33}^{-1}A_{31}\right)^T Q_{33} \\ &- \left(A_{21} - A_{23}A_{33}^{-1}A_{31}\right)^T P_{22} - P_{11}\left(A_{12} - A_{13}A_{33}^{-1}A_{32}\right)\Big] A_{f1}^{-1}. \end{aligned} \tag{4.75}$$

Substituting (4.75) and (4.74) in (4.61), eliminating the terms containing P_{22} by using (4.68) and introducing $L = N + \tilde{A}_{21}^T O + \tilde{A}_{31}^T P_{33} B_3$ converts (4.61) into

$$A_s^T P_{11} + P_{11}A_s + Q_s - LR^{-1}L^T = 0, \tag{4.76}$$

where $\tilde{A}_{21}$, $\tilde{A}_{31}$ and Q_s are given respectively by (4.45), (4.46) and (4.42).

Expression for L can be reframed by substituting N and O, and subsequently using (4.67), (4.74) and (4.75). Replacing P_{22} from the resulting expression by using (4.68), and further replacing H by using (4.69),

$$\begin{aligned} L &= P_{11}B_s + \tilde{Q}_s - LR^{-1}\left(B_3^T P_{33}\tilde{B}_3 + O^T\tilde{B}_2\right) \\ &= N_s - LR^{-1}\left(B_3^T P_{33}\tilde{B}_3 + O^T\tilde{B}_2\right), \quad \text{or,} \\ L &= N_s\left[I + R^{-1}\left(B_3^T P_{33}\tilde{B}_3 + O^T\tilde{B}_2\right)\right]^{-1} = N_s O_s. \end{aligned}$$

Here $\tilde{Q}_s$, $\tilde{B}_2$ and $\tilde{B}_3$ are given respectively by (4.44), (4.47) and (4.48). Expression for O_s^{-1} can further be simplified by expanding O in which (4.67) and (4.68) are inserted, to obtain

$$O_s^{-1} = R^{-1}R_s - R^{-1}\left(\tilde{B}_2^T O + \tilde{B}_3^T P_{33}B_3\right) O_s^{-1}$$

which yields

$$\begin{aligned} O_s &= R_s^{-1}R\left[I + R^{-1}\left(\tilde{B}_2^T O + \tilde{B}_3^T P_{33}B_3\right)\right] \\ &= R_s^{-1}\left(O_s^T\right)^{-1}R \end{aligned}$$

and thereby,

$$O_s R^{-1} O_s^T = R_s^{-1}.$$

This yields $LR^{-1}L^T = N_s R_s^{-1} N_s^T$, with which (4.76) becomes

$$A_s^T P_{11} + P_{11}A_s + Q_s - N_s R_s^{-1} N_s^T = 0. \tag{4.77}$$

Finally, comparison of (4.77) with (4.50) shows that $P_s = P_{11}$.

Chapter 5
Direct Block Diagonalization and Composite Control of Three–Time–Scale Systems

5.1 Introduction

Singular perturbation methods have successfully been used in control applications to deal with multi–time–scale systems, by which the system is decomposed into a 'slow' subsystem and one or more 'boundary layer' or 'fast' subsystems[56]. A system expressed in explicit singularly perturbed form generally has a small parameter ε appearing as a multiplier to the derivative of the 'fast' variables. Here, the system decomposition is achieved by setting $\varepsilon = 0$ and solving for the 'fast' subsystem variables in terms of the 'slow' ones, and then substituting them in the 'slow' subsystem equations[58].

However, in many applications like the AHWR, the scaling parameter ε need not appear explicitly in the model equations. A system model can still have multiple distinct clusters of eigenvalues in spite of not being in an explicit singularly perturbed form. Hence, direct application of singular perturbation techniques to decompose the original system into a 'slow' subsystem and one or more 'boundary layer' subsystems may not be always feasible. Instead, for such systems, direct block diagonalization to decouple it into two or more subsystems could be a more promising approach. In fact, singularly perturbed systems are a class of multi–time–scale systems and the former can be expressed in the latter form[58]. Another benefit of direct block diagonalization is that this technique retains the controllability and observability of individual subsystems of an original high order, controllable and observable system. This is not the case with the application of singular perturbations, where controllability and observability of the original full order system need not guarantee controllability and observability of individual subsystems.

But the literature pertaining to direct applications of multi–time–scale decomposition in not rich. A method for direct block diagonalization of a three–time–scale system is addressed in Naidu [74]. In the following, this technique is revisited and extended to the design of a composite controller.

Consider a general linear discrete time invariant three–time–scale system of order n, represented as

S.R. Shimjith et al.: Modeling and Control of a Large Nuclear Reactor, LNCIS 431, pp. 101–114.
springerlink.com

$$\begin{bmatrix} z_{1_{k+1}} \\ z_{2_{k+1}} \\ z_{3_{k+1}} \end{bmatrix} = \begin{bmatrix} \Phi_{11} & \Phi_{12} & \Phi_{13} \\ \Phi_{21} & \Phi_{22} & \Phi_{23} \\ \Phi_{31} & \Phi_{32} & \Phi_{33} \end{bmatrix} \begin{bmatrix} z_{1_k} \\ z_{2_k} \\ z_{3_k} \end{bmatrix} + \begin{bmatrix} \Gamma_1 \\ \Gamma_2 \\ \Gamma_3 \end{bmatrix} u_k \tag{5.1}$$

where $z_1 \in \Re^{n_1}$, $z_2 \in \Re^{n_2}$ and $z_3 \in \Re^{n_3}$ such that $n_1+n_2+n_3=n$, and $\Phi_{11}, \Phi_{12}, \ldots \Phi_{33}$ and Γ_1, Γ_2 and Γ_3 are corresponding submatrices of the system matrix and input matrix. It is assumed that the system is controllable. Let the eigenspectrum

$$\varphi(\Phi) = (\varphi_1, ..\varphi_{n_1}, \varphi_{n_1+1}, ..\varphi_{n_1+n_2}, \varphi_{n_1+n_2+1}, ..\varphi_n)$$

of the system (5.1) be ordered such that

$$|\varphi_1| > |\varphi_2| > \ldots > |\varphi_{n_1}| \gg |\varphi_{n_1+1}| > \ldots > |\varphi_{n_1+n_2}| \gg |\varphi_{n_1+n_2+1}| \ldots > |\varphi_n| \geq 0.$$

Thus the system possesses three distinct groups of eigenvalues such that n_1 eigenvalues are close to the unit circle and n_2 and n_3 eigenvalues are far and farther from the unit circle towards the origin. Alternatively, the system has n_1 'slow' modes, and n_2 and n_3 'fast' and 'faster' modes.

5.2 Three–Time–Scale Decomposition

Based on the two stage block diagonalization methods presented in [15, 56], a method for block diagonalization and thereby decomposition of a three–time–scale system into 'slow', 'fast 1' and 'fast 2' subsystems is presented in [74]. It achieves the block diagonalization through a three stage transformation. In the first stage, the transformation matrix

$$T_1 = \begin{bmatrix} E_{n_1} & 0 & 0 \\ 0 & E_{n_2} & 0 \\ L_{31} & L_{32} & E_{n_3} \end{bmatrix}$$

is obtained. Here, E_1, E_2 and E_3 are respectively n_1, n_2 and n_3 identity matrices and $n_3 \times n_1$ matrix L_{31} and $n_3 \times n_2$ matrix L_{32} satisfy the non–symmetric algebraic Riccati equations

$$L_{31}\Phi_{11} - L_{31}\Phi_{13}L_{31} + L_{32}\Phi_{21} - L_{32}\Phi_{23}L_{31} - \Phi_{33}L_{31} + \Phi_{31} = 0;$$
$$L_{31}\Phi_{12} - L_{31}\Phi_{13}L_{32} + L_{32}\Phi_{22} - L_{32}\Phi_{23}L_{32} - \Phi_{33}L_{32} + \Phi_{32} = 0.$$

In the second stage, the transformation matrix T_2 given by

$$T_2 = \begin{bmatrix} E_{n_1} & 0 & 0 \\ L_{21} & E_{n_2} & L_{23} \\ 0 & 0 & E_{n_3} \end{bmatrix}$$

is obtained. Here $n_2 \times n_1$ matrix L_{21} and $n_2 \times n_3$ matrix L_{23} satisfy

$$L_{21}(\Phi_{11} - \Phi_{13}L_{31}) - \Phi_{22}L_{21} - L_{21}(\Phi_{12} - \Phi_{13}L_{32})L_{21} + (\Phi_{21} - \Phi_{23}L_{31}) = 0;$$
$$L_{21}\Phi_{13} - L_{21}(\Phi_{12} - \Phi_{13}L_{31})L_{23} - \Phi_{22}L_{23} + \Phi_{23} + L_{23}(\Phi_{33} + L_{31}\Phi_{13} + L_{32}\Phi_{23}) = 0.$$

Finally, the third stage transformation matrix T_3 is obtained as

$$T_3 = \begin{bmatrix} E_{n_1} & L_{12} & L_{13} \\ 0 & E_{n_2} & 0 \\ 0 & 0 & E_{n_3} \end{bmatrix}$$

where $n_1 \times n_2$ matrix L_{12} and $n_1 \times n_3$ matrix L_{13} obey

$$(\Phi_{12} - \Phi_{13}L_{32}) + L_{12}(\Phi_{22} + L_{21}(\Phi_{12} - \Phi_{13}L_{32}))$$
$$+ (\Phi_{11} - \Phi_{13}L_{31} - (\Phi_{12} - \Phi_{13}L_{32})L_{21})L_{12} = 0;$$
$$(\Phi_{11} - \Phi_{13}L_{31} - (\Phi_{12} - \Phi_{13}L_{32})L_{21})L_{13} + \Phi_{13} + L_{13}(\Phi_{33} + L_{31}\Phi_{13} + L_{32}\Phi_{23}) = 0.$$

The submatrices L_{ij} are computed using an iterative procedure described in [56]. The system in (5.1) is then transformed into block diagonal form

$$\begin{bmatrix} z_{s_{k+1}} \\ z_{f1_{k+1}} \\ z_{f2_{k+1}} \end{bmatrix} = \begin{bmatrix} \Phi_s & 0 & 0 \\ 0 & \Phi_{f1} & 0 \\ 0 & 0 & \Phi_{f2} \end{bmatrix} \begin{bmatrix} z_{s_k} \\ z_{f1_k} \\ z_{f2_k} \end{bmatrix} + \begin{bmatrix} \Gamma_s \\ \Gamma_{f1} \\ \Gamma_{f2} \end{bmatrix} u_k, \tag{5.2}$$

through the transformation

$$\left[z_s^T \; z_{f1}^T \; z_{f2}^T \right]^T = T \left[z_1^T \; z_2^T \; z_3^T \right]^T, \tag{5.3}$$

where $T = T_3T_2T_1$. Note that the original n^{th} order system given by (5.1) is decoupled into three subsystems in (5.2), namely, the 'slow' subsystem represented by (Φ_s, Γ_s), the 'fast 1' subsystem represented by (Φ_{f1}, Γ_{f1}) and the 'fast 2' subsystem by (Φ_{f2}, Γ_{f2}), of order n_1, n_2 and n_3 respectively.

5.3 Composite Control

Two stage state feedback controller design for two–time–scale systems, based on the explicitly invertible block diagonalizing transformations is well addressed in literature [72, 82]. First a controller for the fast subsystem is designed. With this, however, the resulting closed loop system is to be again transformed into block diagonal form. Finally, a controller for the 'slow' subsystem of the closed loop system is designed. Here, this method is extended for a three–time–scale system as follows.

Consider the open loop system (5.1) which is equivalently represented in block diagonal form (5.2). For brevity, let $z_d = \left[z_s^T \; z_{f1}^T \; z_{f2}^T \right]^T$. For (5.2), consider

$$u_k = [0 \;\; 0 \;\; F_{f2}] z_{d_k} + u_{2_k}, \tag{5.4}$$

where F_{f2} is the state feedback gain designed for (Φ_{f2}, Γ_{f2}) pair such that eigenvalues of $\Phi_{f2} + \Gamma_{f2} F_{f2}$ are placed at the desired locations. Application of this input to (5.2) results in the closed loop system as

$$z_{d_{k+1}} = \begin{bmatrix} \Phi_s & 0 & \Gamma_s F_{f2} \\ 0 & \Phi_{f1} & \Gamma_{f1} F_{f2} \\ 0 & 0 & \Phi_{f2} + \Gamma_{f2} F_{f2} \end{bmatrix} z_{d_k} + \begin{bmatrix} \Gamma_s \\ \Gamma_{f1} \\ \Gamma_{f2} \end{bmatrix} u_{2_k}.$$

Let $z_{d1} = \left[z_s^T \; \bar{z}_{f1}^T \; z_{f2}^T \right]^T = T_{d1} x_d$, where

$$T_{d1} = \begin{bmatrix} E_1 & 0 & 0 \\ \bar{L}_{21} & E_2 & \bar{L}_{23} \\ 0 & 0 & E_3 \end{bmatrix}, \tag{5.5}$$

and $\bar{L}_{21}$ and $\bar{L}_{23}$ satisfy respectively

$$\bar{L}_{21} \Phi_s - \Phi_{f1} \bar{L}_{21} = 0$$
$$\bar{L}_{23} \left(\Phi_{f2} + \Gamma_{f2} F_{f2} \right) - \Phi_{f1} \bar{L}_{23} + \left(\bar{L}_{21} \Gamma_s F_{f2} + \Gamma_{f1} F_{f2} \right) = 0.$$

Then the following equivalent system is obtained:

$$z_{d1_{k+1}} = \begin{bmatrix} \Phi_s & 0 & \Gamma_s F_{f2} \\ 0 & \Phi_{f1} & 0 \\ 0 & 0 & \Phi_{f2} + \Gamma_{f2} F_{f2} \end{bmatrix} z_{d1_k} + \begin{bmatrix} \Gamma_s \\ \bar{\Gamma}_{f1} \\ \Gamma_{f2} \end{bmatrix} u_{2_k},$$

where

$$\bar{\Gamma}_{f1} = \bar{L}_{21} \Gamma_s + \Gamma_{f1} + \bar{L}_{23} \Gamma_{f2}. \tag{5.6}$$

Since (Φ_{f1}, Γ_{f1}) pair is controllable, $(\Phi_{f1}, \bar{\Gamma}_{f1})$ pair is also controllable [72]. Let F_{f1} be the state feedback gain in

$$u_{2_k} = [0 \;\; F_{f1} \;\; 0] z_{d1_k} + u_{3_k}, \tag{5.7}$$

such that the closed loop system

$$z_{d1_{k+1}} = \begin{bmatrix} \Phi_s & \Gamma_s F_{f1} & \Gamma_s F_{f2} \\ 0 & \Phi_{f1} + \bar{\Gamma}_{f1} F_{f1} & 0 \\ 0 & \Gamma_{f2} F_{f1} & \Phi_{f2} + \Gamma_{f2} F_{f2} \end{bmatrix} z_{d1_k} + \begin{bmatrix} \Gamma_s \\ \bar{\Gamma}_{f1} \\ \Gamma_{f2} \end{bmatrix} u_{3_k} \tag{5.8}$$

has desired stability and transient response properties. Applying a second stage transformation as

$$z_{d2} = \left[\bar{z}_s^T \; \bar{z}_{f1}^T \; z_{f2}^T \right]^T = T_{d2} z_{d1},$$

where

$$T_{d2} = \begin{bmatrix} E_1 & \overline{L}_{12} & \overline{L}_{13} \\ 0 & E_2 & 0 \\ 0 & 0 & E_3 \end{bmatrix}, \tag{5.9}$$

and $\overline{L}_{12}$ and $\overline{L}_{13}$ satisfy

$$\begin{aligned} \overline{L}_{13}\left(\Phi_{f2} + \Gamma_{f2}F_{f2}\right) - \Phi_s\overline{L}_{13} + \Gamma_s F_{f2} &= 0; \\ \overline{L}_{12}\left(\Phi_{f1} + \overline{\Gamma}_{f1}F_{f1}\right) - \Phi_s\overline{L}_{12} + \overline{L}_{13}\Gamma_{f2}F_{f1} &= 0, \end{aligned}$$

it is obtained

$$z_{d2_{k+1}} = \begin{bmatrix} \Phi_s & 0 & 0 \\ 0 & \Phi_{f1} + \overline{\Gamma}_{f1}F_{f1} & 0 \\ 0 & \Gamma_{f2}F_{f1} & \Phi_{f2} + \Gamma_{f2}F_{f2} \end{bmatrix} z_{d2_k} + \begin{bmatrix} \overline{\Gamma}_s \\ \overline{\Gamma}_{f1} \\ \Gamma_{f2} \end{bmatrix} u_{3_k}, \tag{5.10}$$

where $\overline{\Gamma}_s = \Gamma_s + \overline{L}_{12}\overline{\Gamma}_{f1} + \overline{L}_{13}\Gamma_{f2}$. Finally let

$$u_{3_k} = [F_s \;\; 0 \;\; 0] z_{d2_k}, \tag{5.11}$$

where F_s is the controller designed for the $(\Phi_s, \overline{\Gamma}_s)$ pair. Now, from (5.4), (5.7) and (5.11), the composite control $u = u_1 + u_2 + u_3$ is given by

$$u = \left(\left[0 \;\; 0 \;\; F_{f2}\right] + \left[0 \;\; F_{f1} \;\; 0\right] T_{d1} + [F_s \;\; 0 \;\; 0] T_{d2}T_{d1}\right) z_d, \tag{5.12}$$

which turns out to be

$$u = \left[F_s \; F_{f1} \; F_{f2}\right] T_{d2}T_{d1}z_d, \tag{5.13}$$

$$\text{or,} \quad u = \left[F_s \; F_{f1} \; F_{f2}\right] T_{d2}T_{d1}Tz. \tag{5.14}$$

Recall that z is the state vector of the original three–time–scale system given by (5.1).

5.4 Application to the AHWR Model

5.4.1 *Block Diagonalization*

In order to decompose and to design a spatial power controller, the AHWR model with total power feedback given by (3.3) is discretized to obtain

$$\begin{aligned} z_{k+1} &= \Phi z_k + \Gamma_1 \overline{u}_k + \Gamma_2 u_k, \\ y_k &= M z_k; \\ \text{where } \Phi &= e^{(A - B_1 KM)\triangle T}; \end{aligned} \tag{5.15}$$

$$\Gamma_1 = \int_{\tau=0}^{\triangle T} e^{(A-B_1KM)\tau} B_1 d\tau; \text{ and,}$$
$$\Gamma_2 = \int_{\tau=0}^{\triangle T} e^{(A-B_1KM)\tau} B_2 d\tau.$$

Here the sampling interval $\triangle T$ is chosen as 10 s so as to retain the three time scale property in the discrete–time model as well. Three–time–scale decomposition technique described in Sec. 5.2 is applied to the resultant discrete–time model as to decompose the system into slow, fast 1 and fast 2 subsystems of order 38, 35 and 17 respectively, with a structure

$$\begin{bmatrix} z_{s_{k+1}} \\ z_{f1_{k+1}} \\ z_{f2_{k+1}} \end{bmatrix} = \begin{bmatrix} \Phi_s & 0 & 0 \\ 0 & \Phi_{f1} & 0 \\ 0 & 0 & \Phi_{f2} \end{bmatrix} \begin{bmatrix} z_{s_k} \\ z_{f1_k} \\ z_{f2_k} \end{bmatrix} + \begin{bmatrix} \Gamma_{1_s} \\ \Gamma_{1_{f1}} \\ \Gamma_{1_{f2}} \end{bmatrix} u_{1_k} + \begin{bmatrix} \Gamma_{2_s} \\ \Gamma_{2_{f1}} \\ \Gamma_{2_{f2}} \end{bmatrix} u_{2_k}. \quad (5.16)$$

Table 5.1 lists a comparison of the eigenvalues of the original system and the decomposed subsystems, which shows good agreement among the two. It is also observed that the $(\Phi_s, \Gamma_{1_s}, M_s)$, $\left(\Phi_{f1}, \Gamma_{1_{f1}}, M_{f1}\right)$ and $\left(\Phi_{f2}, \Gamma_{1_{f2}}, M_{f2}\right)$ are controllable and observable. In the following, it is attempted to design state feedback regulators for the individual subsystems and augment them to obtain a composite regulator for the original system [99].

5.4.2 Controller Design

It is desired to construct a state feedback controller for the system (6.29), which is approximately decomposed into slow, fast 1 and fast 2 subsystems of order 38, 35 and 17 respectively. To accomplish this, state feedback regulators F_s, F_{f1} and F_{f2} are designed for the individual subsystems so as to place the closed loop poles at the desired locations as listed in Table 5.2, and then a composite regulator was evolved using (5.14). The gain matrix thus designed has elements varying in magnitude from 0 to about 1000. The elements corresponding to the last 17 states are all zeros. It can be observed from Table 5.2 that the closed loop eigenvalues of the original system with the composite regulator is in good agreement with those of the individual subsystems.

5.4.3 Transient Response

The closed loop response of the system with the composite controller is assessed by non–linear simulations. The component in the RR control signals corresponding to the total power control was generated and applied at finer time steps so as to correspond to continuous time control while the spatial control component using the

Table 5.1 Comparison of Original and Subsystem Eigenvalues of the AHWR Model with Total Power Feedback, Discretized with $10s$ sampling interval.

k	ψ_k(orig. system)	ψ_k(subsystem)	k	ψ_k(orig. system)	ψ_k(subsystem)
		Slow Subsystem	50	5.3826×10^{-1}	5.3826×10^{-1}
$1-2$	$1.0009 \pm j2.1819 \times 10^{-4}$	$1.0009 \pm j2.1819 \times 10^{-4}$	51	5.3687×10^{-1}	5.3687×10^{-1}
$3-4$	$1.0008 \pm j3.9896 \times 10^{-4}$	$1.0008 \pm j3.9896 \times 10^{-4}$	52	5.3590×10^{-1}	5.3590×10^{-1}
5	9.9991×10^{-1}	9.9991×10^{-1}	53	5.3549×10^{-1}	5.3549×10^{-1}
6	9.9983×10^{-1}	9.9983×10^{-1}	54	5.3468×10^{-1}	5.3468×10^{-1}
7	9.9980×10^{-1}	9.9980×10^{-1}	55	5.3331×10^{-1}	5.3331×10^{-1}
8	9.9975×10^{-1}	9.9975×10^{-1}	56	5.3316×10^{-1}	5.3316×10^{-1}
$9-10$	$9.9964 \times 10^{-1} \pm j7.6589 \times 10^{-4}$	$9.9964 \times 10^{-1} \pm j7.6589 \times 10^{-4}$	57	3.0992×10^{-1}	3.0992×10^{-1}
11	9.9962×10^{-1}	9.9962×10^{-1}	58	2.2966×10^{-1}	2.2966×10^{-1}
$12-13$	$9.9962 \times 10^{-1} \pm j7.6446 \times 10^{-4}$	$9.9962 \times 10^{-1} \pm j7.6446 \times 10^{-4}$	59	2.2962×10^{-1}	2.2962×10^{-1}
14	9.9962×10^{-1}	9.9962×10^{-1}	60	2.2744×10^{-1}	2.2744×10^{-1}
15	9.9960×10^{-1}	9.9960×10^{-1}	61	2.2651×10^{-1}	2.2651×10^{-1}
16	9.9958×10^{-1}	9.9958×10^{-1}	62	2.1056×10^{-1}	2.1056×10^{-1}
17	9.9958×10^{-1}	9.9958×10^{-1}	63	2.1046×10^{-1}	2.1046×10^{-1}
18	9.9956×10^{-1}	9.9956×10^{-1}	64	2.0883×10^{-1}	2.0883×10^{-1}
19	9.9953×10^{-1}	9.9953×10^{-1}	65	2.0680×10^{-1}	2.0680×10^{-1}
20	9.9951×10^{-1}	9.9951×10^{-1}	66	1.9561×10^{-1}	1.9561×10^{-1}
$21-22$	$9.9935 \times 10^{-1} \pm j5.3075 \times 10^{-4}$	$9.9935 \times 10^{-1} \pm j5.3075 \times 10^{-4}$	67	1.9545×10^{-1}	1.9545×10^{-1}
$23-24$	$9.9934 \times 10^{-1} \pm j5.4926 \times 10^{-4}$	$9.9934 \times 10^{-1} \pm j5.4926 \times 10^{-4}$	68	1.9388×10^{-1}	1.9388×10^{-1}
$25-26$	$9.9927 \times 10^{-1} \pm j3.9290 \times 10^{-4}$	$9.9927 \times 10^{-1} \pm j3.9290 \times 10^{-4}$	69	1.9056×10^{-1}	1.9056×10^{-1}
$27-28$	$9.9923 \times 10^{-1} \pm j2.9906 \times 10^{-4}$	$9.9923 \times 10^{-1} \pm j2.9906 \times 10^{-4}$	70	1.6469×10^{-1}	1.6469×10^{-1}
29	9.9859×10^{-1}	9.9859×10^{-1}	71	1.6449×10^{-1}	1.6449×10^{-1}
30	9.9854×10^{-1}	9.9854×10^{-1}	72	1.6330×10^{-1}	1.6330×10^{-1}
31	9.9843×10^{-1}	9.9843×10^{-1}	73	1.5884×10^{-1}	1.5884×10^{-1}
32	9.9835×10^{-1}	9.9835×10^{-1}			Fast 2 Subsystem
33	9.9834×10^{-1}	9.9834×10^{-1}	74	0.0	0.0
34	9.9827×10^{-1}	9.9827×10^{-1}	75	0.0	0.0
35	9.9812×10^{-1}	9.9812×10^{-1}	76	0.0	0.0
36	9.9811×10^{-1}	9.9811×10^{-1}	77	0.0	0.0
$37-38$	9.9553×10^{-1}	9.9512×10^{-1}	78	0.0	0.0
		Fast 1 Subsystem	79	0.0	0.0
39	8.4158×10^{-1}	8.4235×10^{-1}	80	0.0	0.0
40	7.5279×10^{-1}	7.5279×10^{-1}	81	0.0	0.0
41	6.0077×10^{-1}	6.0077×10^{-1}	82	0.0	0.0
42	5.9954×10^{-1}	5.9954×10^{-1}	83	0.0	0.0
43	5.6141×10^{-1}	5.6141×10^{-1}	84	0.0	0.0
44	5.6050×10^{-1}	5.6050×10^{-1}	85	0.0	0.0
45	5.5042×10^{-1}	5.5042×10^{-1}	86	0.0	0.0
46	5.5033×10^{-1}	5.5033×10^{-1}	87	0.0	0.0
47	5.4693×10^{-1}	5.4693×10^{-1}	88	0.0	0.0
48	5.4530×10^{-1}	5.4530×10^{-1}	89	0.0	0.0
49	5.3876×10^{-1}	5.3876×10^{-1}	90	0.0	0.0

composite controller was generated and superimposed on the total power control signal every $10s$. The system was initially assumed to be at full power steady state conditions, with all RRs at their equilibrium positions. A transient identical to that described in Sec. 3.4 was introduced. Fig. 5.1 depicts the control signals generated by the spatial power controller (5.14), and Fig. 5.2 depicts the corresponding RR positions. It is observed that the disturbances in nodal power variations, exit qualities and quadrant core powers are stabilized within about 150 s, as shown in Fig. 5.3, Fig. 5.4 and Fig. 5.5 respectively. From Fig. 5.6, it may be noticed that the total power underwent an overshoot by about 0.5% which is stabilized rather instantaneously, and thereafter remaining constant. No tilts were also observed during a

Table 5.2 Comparison of Closed Loop Eigenvalues of the Original and Decomposed Subsystems.

Sl.No.	Closed Loop Eigenvalues		Sl.No.	Closed Loop Eigenvalues	
	Slow Subsyst.	Orig. syst. (Composite Regulator)		Fast 1 Subsyst. (contd.)	Orig. Syst. (Composite Regulator)
1	0.99998	0.99983	46	0.53825	0.53821
2	0.99965	0.99962	47	0.53686	0.53612
3	0.99965	0.99962	48	0.53589	0.53576
4	0.99962	0.99962	49	0.53548	0.53499
5	0.99962	0.99962	50	0.53467	0.53459
6	0.99962	0.99961	51	0.53330	0.53318
7	0.99962	0.99961	52	0.53316	0.53308
8	0.99960	0.99960	53	0.31000	0.31010
9	0.99958	0.99958	54	0.22965	0.23321
10	0.99958	0.99958	55	0.22961	0.23032
11	0.99956	0.99956	56	0.22736	0.22966
12	0.99953	0.99953	57	0.22695	0.22737
13	0.99951	0.99951	58	0.21055	0.21065
14	0.99935	0.99941	59	0.21043	0.20998
15	0.99935	0.99941	60	0.20874	0.20723
16	0.99934	0.99935	61	0.20726	0.20626
17	0.99934	0.99935	62	0.20364	0.20417
18	0.99927	0.99934	63	0.20347	0.19941
19	0.99927	0.99934	64	0.20282	0.19906
20	0.99923	0.99927	65	0.19560	0.19501
21	0.99923	0.99927	66	0.19543	0.19384
22	0.99919	0.99923	67	0.19377	0.19299
23	0.99919	0.99923	68	0.19190	0.19299
24	0.99912	0.99911	69	0.16468	0.16468
25	0.99912	0.99911	70	0.16446	0.16363
26	0.99859	0.99859	71	0.16322	0.16298
27	0.99854	0.99854	72	0.16075	0.16012
28	0.99843	0.99843	73	0.02402	0.02325
29	0.99835	0.99835		Fast 2 Subsystem	
30	0.99833	0.99834	74	0.00000	0.00000
31	0.99827	0.99827	75	0.00000	0.00000
32	0.99812	0.99812	76	0.00000	0.00000
33	0.99811	0.99811	77	0.00000	0.00000
34	0.89133	0.93027	78	0.00000	0.00000
35	0.85322	0.93027	79	0.00000	0.00000
36	0.83876	0.83551	80	0.00000	0.00000
37	0.83215	0.83551	81	0.00000	0.00000
38	0.83110	0.83148	82	0.00000	0.00000
	Fast 1 Subsystem		83	0.00000	0.00000
39	0.56129	0.56112	84	0.00000	0.00000
40	0.56029	0.56001	85	0.00000	0.00000
41	0.55042	0.55035	86	0.00000	0.00000
42	0.55029	0.54951	87	0.00000	0.00000
43	0.54688	0.54548	88	0.00000	0.00000
44	0.54530	0.54491	89	0.00000	0.00000
45	0.53875	0.53825	90	0.00000	0.00000

long term simulation. The response of the controller could be noticed to be faster than that of the controller obtained in the previous chapter, and comparable to that obtained with the output feedback controller.

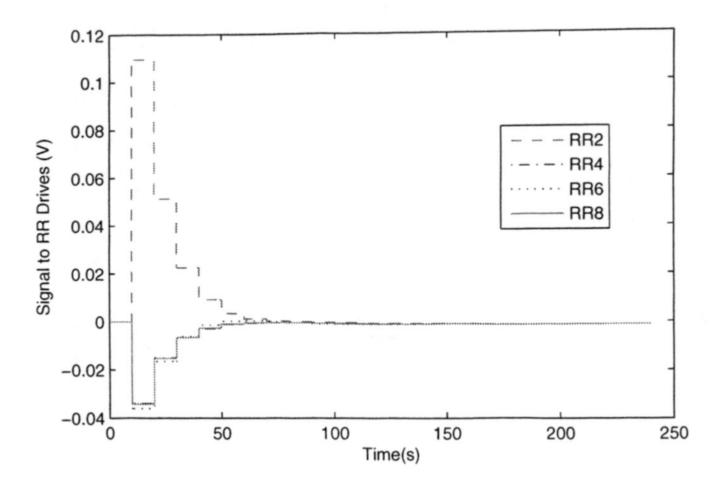

Fig. 5.1 Control Signals Generated by Spatial Controller.

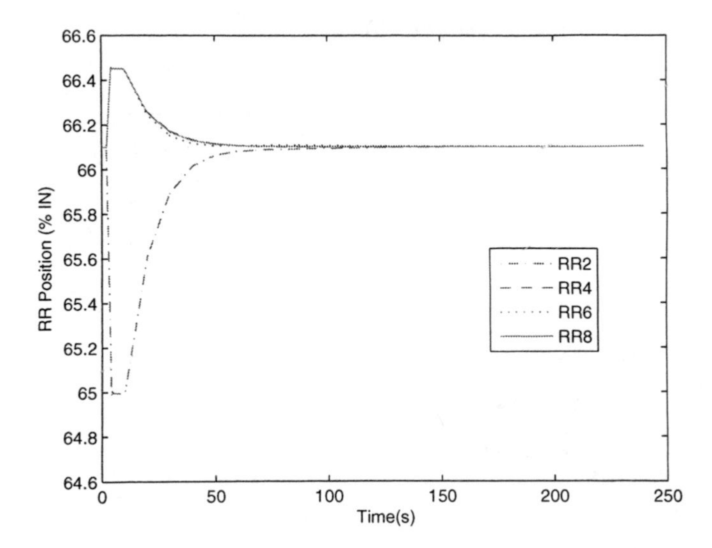

Fig. 5.2 Variations in RR Positions during the Transient.

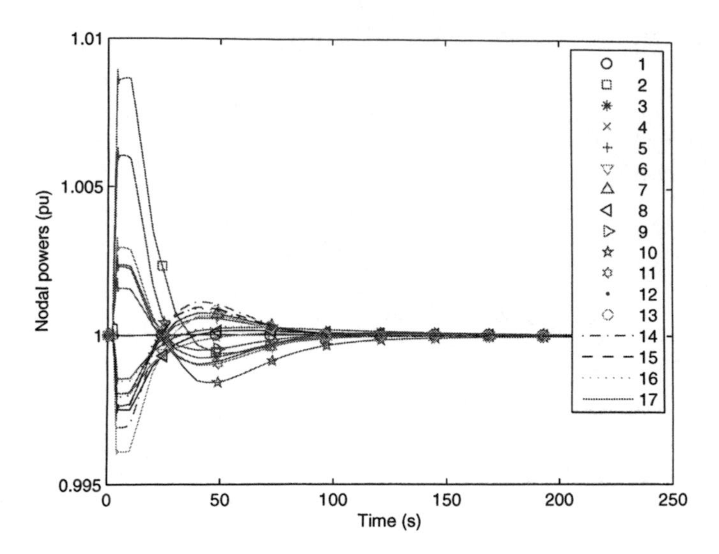

Fig. 5.3 Nodal Power variations during the Transient.

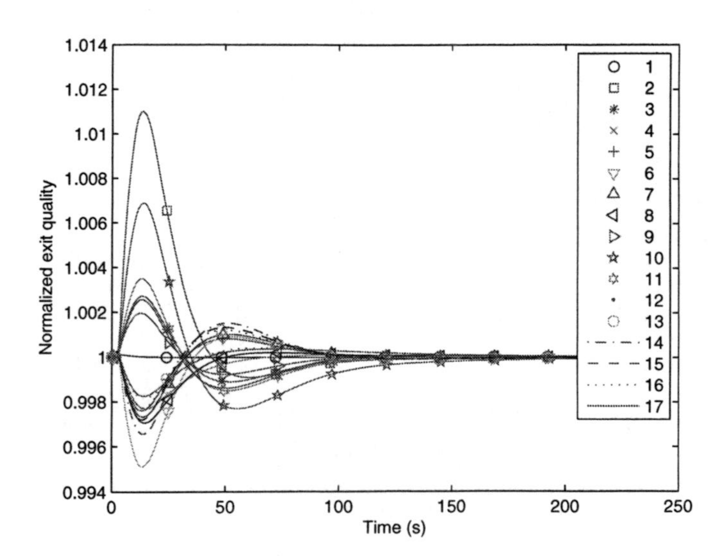

Fig. 5.4 Variations in Exit Qualities during the Transient.

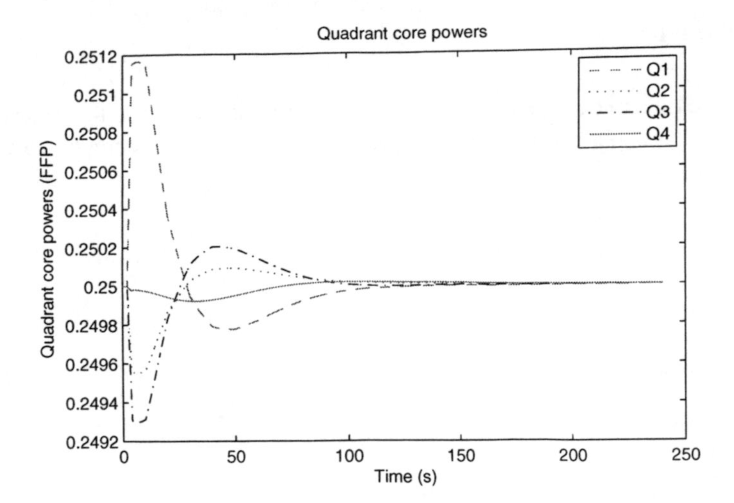

Fig. 5.5 Quadrant Core Power variations during the Transient.

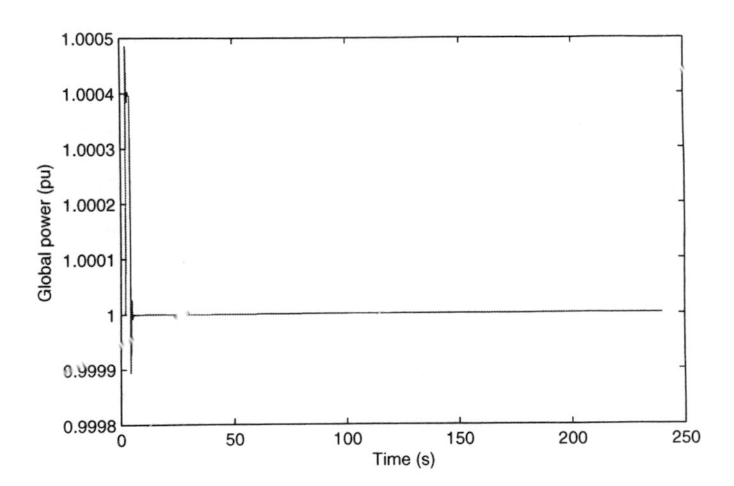

Fig. 5.6 Variations in Total Power during the Transient.

5.5 Robustness

To assess the robustness of the controller to the disturbances in feed flow, a transient identical to that explained in Sec. 3.4.1 was simulated. Fig. 5.7 depicts the change

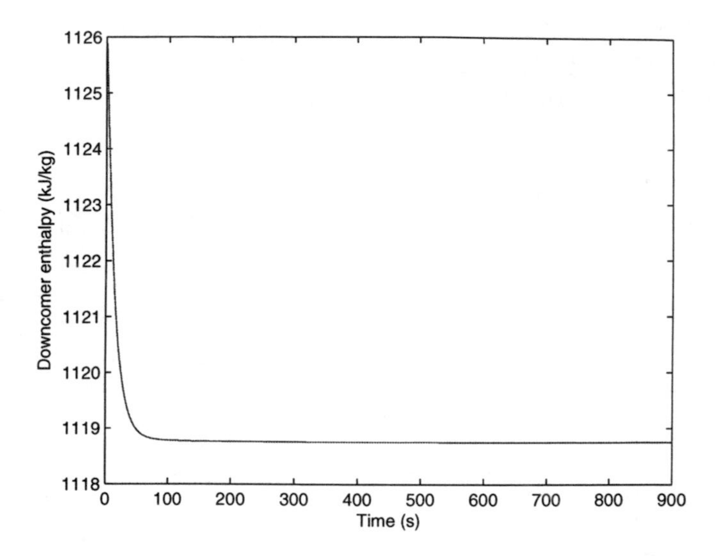

Fig. 5.7 Change in Downcomer Enthalpy due to Step Change in Feed Flow.

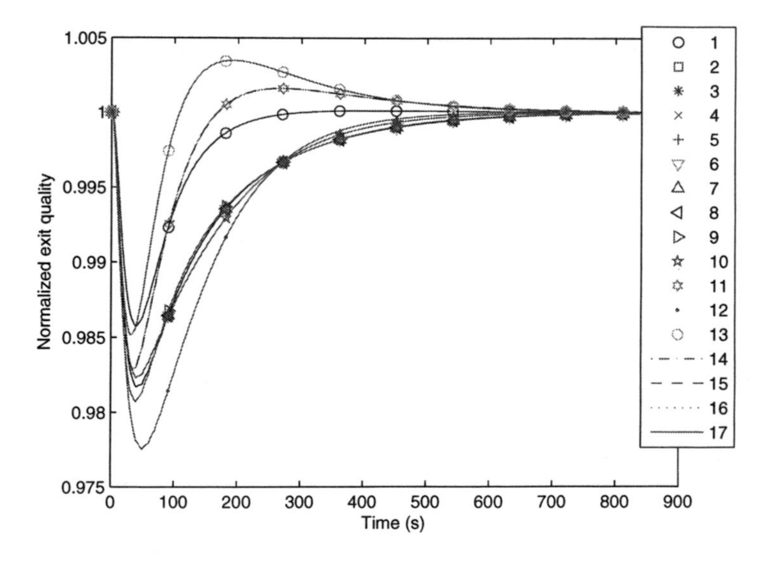

Fig. 5.8 Exit Quality Variations subsequent to Step Change in Feed Flow.

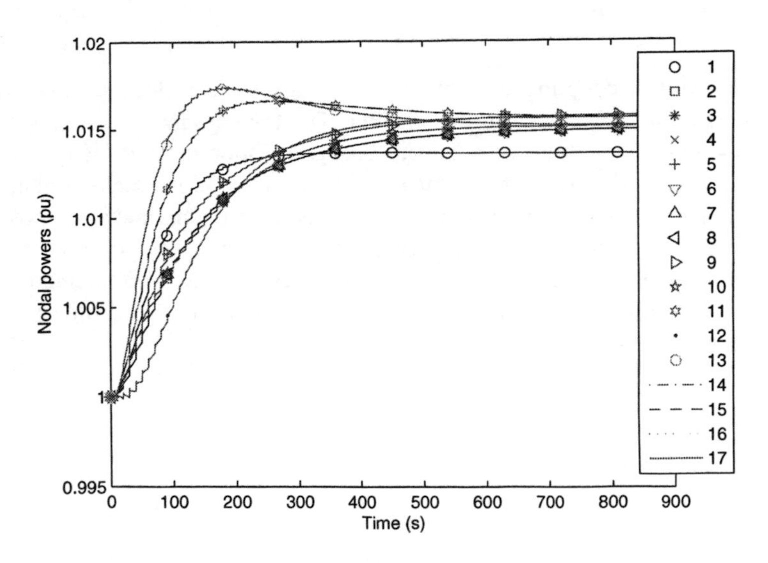

Fig. 5.9 Nodal Power Variations subsequent to Step Change in Feed Flow.

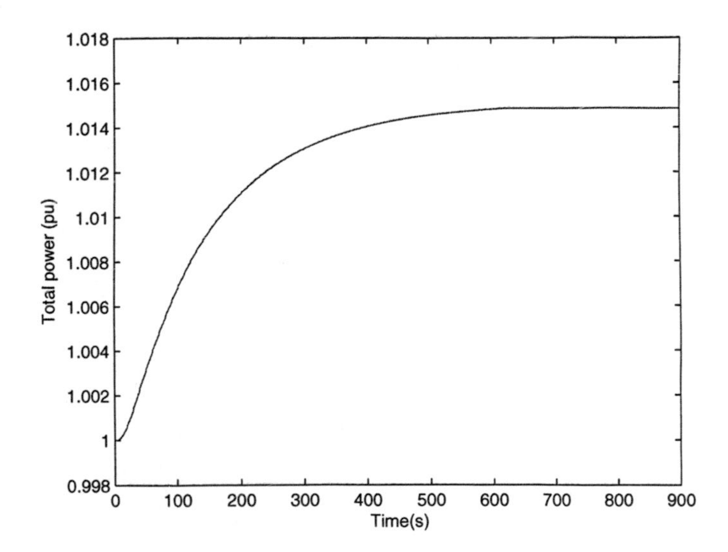

Fig. 5.10 Variations in Total Power during Step Change in Feed Flow.

in downcomer enthalpy subsequent to a 5% step change in the feed water flow rate, and Fig. 5.8 shows the corresponding changes in core exit qualities. Nodal powers and total reactor power underwent disturbances as shown in Fig. 5.9 and Fig. 5.10 respectively, from which it may be noticed that the controller demonstrates a stable response during the transient.

5.6 Conclusion

A method for decoupling of a three–time–scale system through direct block–diagonalization is discussed in this chapter. The transformation matrices required for block–diagonalization are obtained through solution of a set of Riccati equations. Controller design problem for the original three–time–scale system is cast into three smaller order subsystem controller design problems, and a composite controller has been derived from these subsystem controllers. The proposed technique is applied to the spatial control problem of the AHWR, and a stabilizing controller is obtained. Validity of the controller is demonstrated through dynamic simulations of the original non–linear equations.

Chapter 6
Design of Fast Output Sampling Controller for Three–Time–Scale Systems

6.1 Introduction

The practical implementation of a state feedback controller demands a state observer of large order. In this context, an attractive alternative to the state feedback would be an output feedback controller. In the recent past, Fast Output Sampling (FOS) control technique, a method which utilizes only the system output for feedback, has emerged as a promising candidate for control of sampled data control systems[36, 109, 123]. Major advantage of this technique is its ability to guarantee arbitrary symmetric pole placement, thereby yielding a closed–loop stable system. Large number of applications of FOS control technique can be found especially in the areas of simultaneous stabilization, robust control, sliding mode control, nuclear reactor control *etc.* [9, 85, 121, 122].

The FOS is a control technique of non–dynamic output feedback capable of assigning arbitrary dynamical characteristics to the closed loop system. Consider a plant described by linear model of the form

$$\dot{z} = Az + Bu, \tag{6.1}$$

$$y = Mz, \tag{6.2}$$

where (A,B) and (A,M) are assumed to be controllable and observable respectively. Let τ be a sampling interval during which the control input u is maintained constant, and let $\triangle$ be a smaller interval such that $\triangle = \tau/N$. The output measurements are taken at time instant $t = l\triangle$ where $l = 0, 1, \cdots$. The control signal $u(t)$, which is applied during the interval $k\tau < t < (k+1)\tau$, is then constructed as a linear combination of the last N output observations. Here $N \geq \nu$, the observability index of the system.

Definition: Observability Index: *Given an observable pair* $(A,M) \in \Re^{n\times n} \times \Re^{q\times n}$, *and* $rank(M) = q$, *the observability index of the system with respect to any particular row* m_i *of* M *is the minimum value of* ν_i *such that* $m_i A^{\nu_i}$ *is dependent on the rows before it in the following series*

S.R. Shimjith et al.: Modeling and Control of a Large Nuclear Reactor, LNCIS 431, pp. 115–130.
springerlink.com

$$[m_1, m_2, \cdots, m_q, m_1 A, m_2 A, \cdots, m_q A, \cdots, m_1 A^{\nu_i}, \cdots, m_i A^{\nu_i}, \cdots].$$

The observability index of the entire system is defined as $\nu = max(\nu_i)$.

In fast output sampling, the control signal to the system (6.1) is generated as

$$u(t) = [L_0 \; L_1 \; \ldots \; L_{N-1}] \begin{bmatrix} y(k\tau - \tau) \\ y(k\tau - \tau + \triangle) \\ \vdots \\ y(k\tau - \triangle) \end{bmatrix} = Ly_k. \tag{6.3}$$

It may be noticed that the control signal is generated at a rate $1/\tau$, whereas the outputs are sampled at N times faster rate $1/\triangle$.

Let $(\Phi_\tau, \Gamma_\tau, M)$ denote the system (A, B, M) sampled at the rate $1/\tau$, *i.e.*, $\Phi = e^{A\tau}$, $\Gamma_\tau = \int_0^\tau e^{As} B ds$. Also let $(\Phi_\triangle, \Gamma_\triangle, M)$ denote the same system sampled at the rate $1/\triangle$. Consider the discrete time systems having at $t = k\tau$ the input $u_k = u(k\tau)$, states $z_k = z(k\tau)$ and the output y_k. Then we have

$$z_{k+1} = \Phi_\tau z_k + \Gamma_\tau u_k, \tag{6.4}$$

$$y_{k+1} = M_0 z_k + D_0 u_k, \tag{6.5}$$

where

$$M_0 = \begin{bmatrix} M \\ M\Phi_\triangle \\ M\Phi_\triangle^2 \\ \vdots \\ M\Phi_\triangle^{N-1} \end{bmatrix} \quad \text{and} \quad D_0 = \begin{bmatrix} 0 \\ M\Gamma_\triangle \\ M\Phi_\triangle\Gamma_\triangle + M\Gamma_\triangle \\ \vdots \\ M\sum_{i=0}^{N-2} \Phi_\triangle^i \Gamma_\triangle \end{bmatrix}. \tag{6.6}$$

Let $u_k = Fz_k$ be a state feedback control law such that the closed loop system $(\Phi_\tau + \Gamma_\tau F)$ possesses desired response characteristics and no eigenvalues at origin. For this state feedback one can define the fictitious measurement matrix

$$\tilde{M}(F, n) = (M_0 + D_0 F)(\Phi_\tau + \Gamma_\tau F)^{-1} \tag{6.7}$$

which satisfies the fictitious measurement equation

$$y_k = \tilde{M} z_k. \tag{6.8}$$

Then the feedback law (6.3) can be interpreted as static output feedback $u_k = Ly_k$ where L satisfies

$$L\tilde{M} = F. \tag{6.9}$$

At time $t = 0$, the control signal $u(t) = u_0$ for $0 < t \leq \tau$, cannot be computed directly as the output measurements are not available for $t < 0$. If the initial state z_0 is known one can take $u_0 = Fz_0$. If z_0 is unknown but estimated with error $\triangle z_0$, the value of

u_0 will differ by $\triangle u_0 = F\triangle z_0$ from the control signal which would be applied if the initial state was known. This error will propagate through the closed loop response of the system. However, the closed loop dynamics is governed by

$$\begin{bmatrix} z_{k+1} \\ \triangle u_{k+1} \end{bmatrix} = \begin{bmatrix} \Phi_\tau + \Gamma_\tau F & \Gamma_\tau \\ 0 & LD_0 - F\Gamma_\tau \end{bmatrix} \begin{bmatrix} z_k \\ \triangle u_k \end{bmatrix} \tag{6.10}$$

where $\triangle u_k = u_k - F z_k$[122]. Thus we have the eigenvalues of the closed loop system under the fast output sampling law (6.3) as a union of those of $\Phi_\tau + \Gamma_\tau F$ which is stable by design, and those of $LD_0 - F\Gamma_\tau$. If F is designed such that the latter is also stable, the error will decay down to zero.

In most of the cases of designing a fast output sampling controller, a unique solution for (6.3) may not exist. This demands the solution of (6.3) to be optimal rather than an exact one. Several methods have been postulated for computation of the gain L. Tiwari *et.al.*[116] proposed a methodology for computation of the gain matrix for two–time scale systems, which eliminates the associated ill–conditioning issues. A formulation based on LMI approach has been proposed in [109] for design of fast output sampling controller for spatial control of a large PHWR. An LMI approach based design of robust fast output sampling controller for multimodel systems has been proposed in [76]. A modified approach based on multirate output feedback for design of an exact FOS controller, which does not employ the LMI approach and hence suitable for single plants only, has also been developed in recent times [44]. The advantage of this improved version is its capability to realize any state based controller rather than just the static gain state feedback.

It is established that with the FOS gain L, the closed loop system poles are placed at the locations specified by the state feedback gain F. It may be observed that the FOS gain L contains only constant values and hence it is easier to implement. However, it is well known that the straight–forward application of such a controller design method to systems exhibiting multiple time scales, like the AHWR model does, is susceptible to numerical ill–conditioning and stiffness problems. Conversely, the presence of time scales can effectively be exploited for achieving better computational efficiency in case of the design of controller based on fast output sampling principle. Derivation of a fast output sampling controller for two–time–scale systems is presented in [116]. This approach can be extended for design of FOS controller for a three–time–scale system by decomposition into three subsystem problems. On the basis of the approach, systematic procedure for decomposition of the high order FOS design problem into three lower order problems, one for each of the subsystems, can be derived. These smaller order problems can then be separately solved and the results thus obtained for the subsystems can be combined to obtain the solution of the original design problem, as discussed in the subsequent sections.

6.2 FOS for Three–Time–Scale Systems

Consider a linear time invariant system possessing three–time–scale property, i.e., it has n_1 slow modes, n_2 fast modes and n_2 fastest modes. Then, using the technique described in Sec. 5.2, it can be represented in standard block diagonal form as

$$\begin{bmatrix} \dot{z}_s \\ \dot{z}_{f1} \\ \dot{z}_{f2} \end{bmatrix} = \begin{bmatrix} A_s & 0 & 0 \\ 0 & A_{f1} & 0 \\ 0 & 0 & A_{f2} \end{bmatrix} \begin{bmatrix} z_s \\ z_{f1} \\ z_{f2} \end{bmatrix} + \begin{bmatrix} B_s \\ B_{f1} \\ B_{f2} \end{bmatrix} u, \tag{6.11}$$
$$y = \begin{bmatrix} M_s \; M_{f1} \; M_{f2} \end{bmatrix} \begin{bmatrix} z_s^T \; z_{f1}^T \; z_{f2}^T \end{bmatrix}^T,$$

where $z_s \in \Re^{n1}$, $z_{f1} \in \Re^{n2}$, $z_{f2} \in \Re^{n3}$, $u \in \Re^m$ and $y \in \Re^p$. Now the discrete–time system corresponding to this model for sampling period τ becomes

$$\begin{bmatrix} z_{s,k+1} \\ z_{f1,k+1} \\ z_{f2,k+1} \end{bmatrix} = \begin{bmatrix} \Phi_{\tau_s} & 0 & 0 \\ 0 & \Phi_{\tau_{f1}} & 0 \\ 0 & 0 & \Phi_{\tau_{f2}} \end{bmatrix} \begin{bmatrix} z_{s,k} \\ z_{f1,k} \\ z_{f2,k} \end{bmatrix} + \begin{bmatrix} \Gamma_{\tau_s} \\ \Gamma_{\tau_{f1}} \\ \Gamma_{\tau_{f2}} \end{bmatrix} u_k, \tag{6.12}$$
$$y_k = \begin{bmatrix} M_s \; M_{f1} \; M_{f2} \end{bmatrix} \begin{bmatrix} z_{s,k}^T \; z_{f1,k}^T \; z_{f2,k}^T \end{bmatrix}^T,$$

and that for sampling interval $\triangle$ becomes

$$\begin{bmatrix} z_{s,k+1} \\ z_{f1,k+1} \\ z_{f2,k+1} \end{bmatrix} = \begin{bmatrix} \Phi_{\triangle_s} & 0 & 0 \\ 0 & \Phi_{\triangle_{f1}} & 0 \\ 0 & 0 & \Phi_{\triangle_{f2}} \end{bmatrix} \begin{bmatrix} z_{s,k} \\ z_{f1,k} \\ z_{f2,k} \end{bmatrix} + \begin{bmatrix} \Gamma_{\triangle_s} \\ \Gamma_{\triangle_{f1}} \\ \Gamma_{\triangle_{f2}} \end{bmatrix} u_k, \tag{6.13}$$
$$y_k = \begin{bmatrix} M_s \; M_{f1} \; M_{f2} \end{bmatrix} \begin{bmatrix} z_{s,k}^T \; z_{f1,k}^T \; z_{f2,k}^T \end{bmatrix}^T.$$

The discretized systems (6.12) and (6.13) are also in block diagonal form, and would also possess three–time–scale property for suitably selected sampling interval $\triangle$. Moreover, it is easy to comprehend that the eigenvalues of Φ_{τ_s}, $\Phi_{\tau_{f1}}$ and $\Phi_{\tau_{f2}}$ are larger in magnitude than those of $\Phi_{\triangle_s}$, $\Phi_{\triangle_{f1}}$ and $\Phi_{\triangle_{f2}}$, repsctively taken in that order.

Let F_s, F_{f1} and F_{f2} be the state feedback gains designed for the slow, fast 1 and fast 2 subsystems in (6.12), such that

$$|\varphi_i(\Phi_{\tau_s} + \Gamma_{\tau_s} F_s)| < 1, \; i \in [1, n_1], \tag{6.14}$$
$$|\varphi_i\left(\Phi_{\tau_{f1}} + \Gamma_{\tau_{f1}} F_{f1}\right)| < 1, \; i \in [1, n_2], \tag{6.15}$$
$$|\varphi_i\left(\Phi_{\tau_{f2}} + \Gamma_{\tau_{f2}} F_{f2}\right)| < 1, \; i \in [1, n_3], \tag{6.16}$$
$$\min|\varphi_i(\Phi_{\tau_s} + \Gamma_{\tau_s} F_s)| \gg \max|\varphi_j\left(\Phi_{\tau_{f1}} + \Gamma_{\tau_{f1}} F_{f1}\right)|, \tag{6.17}$$
$$\min|\varphi_i\left(\Phi_{\tau_{f1}} + \Gamma_{\tau_{f1}} F_{f1}\right)| \gg \max|\varphi_j\left(\Phi_{\tau_{f2}} + \Gamma_{\tau_{f2}} F_{f2}\right)|. \tag{6.18}$$

If the slow, fast 1 and fast 2 subsystems are controllable, feedback gains F_s, F_{f1} and F_{f2} satisfying (6.14)–(6.18) exist. Then for the system (6.12), the state feedback gain will be given by (5.13) as

$$F = \left[F_1\; F_2\; F_3\right] = \left[F_s\; F_{f1}\; F_{f2}\right] T_{d2} T_{d1}. \tag{6.19}$$

From (6.6),

$$M_0 = \begin{bmatrix} M_s & M_{f1} & M_{f2} \\ M_s \Phi_{\triangle_s} & M_{f1} \Phi_{\triangle_{f1}} & M_{f2} \Phi_{\triangle_{f2}} \\ \vdots & \vdots & \vdots \\ M_s \Phi_{\triangle_s}^{N-1} & M_{f1} \Phi_{\triangle_{f1}}^{N-1} & M_{f2} \Phi_{\triangle_{f2}}^{N-1} \end{bmatrix}, \tag{6.20}$$

$$D_0 = \begin{bmatrix} 0 \\ M_s \Gamma_{\triangle_s} + M_{f1} \Gamma_{\triangle_{f1}} + M_{f2} \Gamma_{\triangle_{f2}} \\ \vdots \\ M_s \sum_{j=0}^{N-2} \Phi_{\triangle_s}^{j} \Gamma_{\triangle_s} + M_{f1} \sum_{j=0}^{N-2} \Phi_{\triangle_{f1}}^{j} \Gamma_{\triangle_{f1}} + M_{f2} \sum_{j=0}^{N-2} \Phi_{\triangle_{f2}}^{j} \Gamma_{\triangle_{f2}} \end{bmatrix} \tag{6.21}$$

Hence, it follows that

$$M_0 + D_0 F = \begin{bmatrix} M_{s_1} & M_{f1_1} & M_{f2_1} \\ M_{s_2} & M_{f1_2} & M_{f2_2} \\ M_{s_3} & M_{f1_3} & M_{f2_3} \end{bmatrix}, \tag{6.22}$$

where the sub–blocks are as given in Table 6.1.

Here $N \geq \nu_{f2} + \nu_{f1} + \nu_s$, where ν_{f2}, ν_{f1} and ν_s are respectively the observability indices of fast 2, fast 1 and slow subsystems. Now consider that the FOS gain matrix L is partitioned into three matrices of respective dimensions $m \times \nu_{f2} p$, $m \times \nu_{f1} p$ and $m \times \left(N - \nu_{f2} - \nu_{f1}\right) p$ as

$$L = \left[L_{f2}\; L_{f1}\; L_s\right]. \tag{6.23}$$

From (6.7) and (6.9) it follows that

$$L(M_0 + D_0 F) = F\left(\Phi_\tau + \Gamma_\tau F\right). \tag{6.24}$$

Let

$$F\left(\Phi_\tau + \Gamma_\tau F\right) = \left[\tilde{F}_1\; \tilde{F}_2\; \tilde{F}_3\right]; \tag{6.25}$$

$$\text{where } \tilde{F}_1 = F_1\left(\Phi_{\tau_s} + \Gamma_{\tau_s} F_1\right) + F_2 \Gamma_{\tau_{f1}} F_1 + F_3 \Gamma_{\tau_{f3}} F_1, \tag{6.26}$$

$$\tilde{F}_2 = F_1 \Gamma_{\tau_s} F_2 + F_2\left(\Phi_{\tau_{f1}} + \Gamma_{\tau_{f1}} F_2\right) + F_3 \Gamma_{\tau_{f3}} F_2, \tag{6.27}$$

$$\tilde{F}_3 = F_1 \Gamma_{\tau_s} F_3 + F_2 \Gamma_{\tau_{f1}} F_3 + F_3\left(\Phi_{\tau_{f2}} + \Gamma_{\tau_{f3}} F_3\right). \tag{6.28}$$

Table 6.1 Sub–blocks of the matrix $(M_0 + D_0F)$ in (6.22).

$$M_{s_1} = \begin{bmatrix} M_s \\ M_s\Phi_{\Delta_s} + \left(M_s\Gamma_{\Delta_s} + M_{f1}\Gamma_{\Delta_{f1}} + M_{f2}\Gamma_{\Delta_{f2}}\right)F_1 \\ \vdots \\ M_s\Phi_{\Delta_s}^{\nu_{f2}-1} + \left(M_s\Sigma_{j=0}^{\nu_{f2}-2}\Phi_{\Delta_s}^{j}\Gamma_{\Delta_s} + M_{f1}\Sigma_{j=0}^{\nu_{f2}-2}\Phi_{\Delta_{f1}}^{j}\Gamma_{\Delta_{f1}} + M_{f2}\Sigma_{j=0}^{\nu_{f2}-2}\Phi_{\Delta_{f2}}^{j}\Gamma_{\Delta_{f2}}\right)F_1 \end{bmatrix},$$

$$M_{s_2} = \begin{bmatrix} M_s\Phi_{\Delta_s}^{\nu_{f2}} + \left(M_s\Sigma_{j=0}^{\nu_{f2}-1}\Phi_{\Delta_s}^{j}\Gamma_{\Delta_s} + M_{f1}\Sigma_{j=0}^{\nu_{f2}-1}\Phi_{\Delta_{f1}}^{j}\Gamma_{\Delta_{f1}} + M_{f2}\Sigma_{j=0}^{\nu_{f2}-1}\Phi_{\Delta_{f2}}^{j}\Gamma_{\Delta_{f2}}\right)F_1 \\ \vdots \\ M_s\Phi_{\Delta_s}^{\nu_{f2}+\nu_{f1}-1} + \left(M_s\Sigma_{j=0}^{\nu_{f2}+\nu_{f1}-2}\Phi_{\Delta_s}^{j}\Gamma_{\Delta_s} + M_{f1}\Sigma_{j=0}^{\nu_{f2}+\nu_{f1}-2}\Phi_{\Delta_{f1}}^{j}\Gamma_{\Delta_{f1}} + M_{f2}\Sigma_{j=0}^{\nu_{f2}+\nu_{f1}-2}\Phi_{\Delta_{f2}}^{j}\Gamma_{\Delta_{f2}}\right)F_1 \end{bmatrix},$$

$$M_{s_3} = \begin{bmatrix} M_s\Phi_{\Delta_s}^{\nu_{f2}+\nu_{f1}} + \left(M_s\Sigma_{j=0}^{\nu_{f2}+\nu_{f1}-1}\Phi_{\Delta_s}^{j}\Gamma_{\Delta_s} + M_{f1}\Sigma_{j=0}^{\nu_{f2}+\nu_{f1}-1}\Phi_{\Delta_{f1}}^{j}\Gamma_{\Delta_{f1}} + M_{f2}\Sigma_{j=0}^{\nu_{f2}+\nu_{f1}-1}\Phi_{\Delta_{f2}}^{j}\Gamma_{\Delta_{f2}}\right)F_1 \\ \vdots \\ M_s\Phi_{\Delta_s}^{N-1} + \left(M_s\Sigma_{j=0}^{N-2}\Phi_{\Delta_s}^{j}\Gamma_{\Delta_s} + M_{f1}\Sigma_{j=0}^{N-2}\Phi_{\Delta_{f1}}^{j}\Gamma_{\Delta_{f1}} + M_{f2}\Sigma_{j=0}^{N-2}\Phi_{\Delta_{f2}}^{j}\Gamma_{\Delta_{f2}}\right)F_1 \end{bmatrix},$$

$$M_{f1_1} = \begin{bmatrix} M_{f1} \\ M_{f1}\Phi_{\Delta_{f1}} + \left(M_s\Gamma_{\Delta_s} + M_{f1}\Gamma_{\Delta_{f1}} + M_{f2}\Gamma_{\Delta_{f2}}\right)F_2 \\ \vdots \\ M_{f1}\Phi_{\Delta_{f1}}^{\nu_{f2}-1} + \left(M_s\Sigma_{j=0}^{\nu_{f2}-2}\Phi_{\Delta_s}^{j}\Gamma_{\Delta_s} + M_{f1}\Sigma_{j=0}^{\nu_{f2}-2}\Phi_{\Delta_{f1}}^{j}\Gamma_{\Delta_{f1}} + M_{f2}\Sigma_{j=0}^{\nu_{f2}-2}\Phi_{\Delta_{f2}}^{j}\Gamma_{\Delta_{f2}}\right)F_2 \end{bmatrix},$$

$$M_{f1_2} = \begin{bmatrix} M_{f1}\Phi_{\Delta_{f1}}^{\nu_{f2}} + \left(M_s\Sigma_{j=0}^{\nu_{f2}-1}\Phi_{\Delta_s}^{j}\Gamma_{\Delta_s} + M_{f1}\Sigma_{j=0}^{\nu_{f2}-1}\Phi_{\Delta_{f1}}^{j}\Gamma_{\Delta_{f1}} + M_{f2}\Sigma_{j=0}^{\nu_{f2}-1}\Phi_{\Delta_{f2}}^{j}\Gamma_{\Delta_{f2}}\right)F_2 \\ \vdots \\ M_{f1}\Phi_{\Delta_{f1}}^{\nu_{f2}+\nu_{f1}-1} + \left(M_s\Sigma_{j=0}^{\nu_{f2}+\nu_{f1}-2}\Phi_{\Delta_s}^{j}\Gamma_{\Delta_s} + M_{f1}\Sigma_{j=0}^{\nu_{f2}+\nu_{f1}-2}\Phi_{\Delta_{f1}}^{j}\Gamma_{\Delta_{f1}} + M_{f2}\Sigma_{j=0}^{\nu_{f2}+\nu_{f1}-2}\Phi_{\Delta_{f2}}^{j}\Gamma_{\Delta_{f2}}\right)F_2 \end{bmatrix},$$

$$M_{f1_3} = \begin{bmatrix} M_{f1}\Phi_{\Delta_{f1}}^{\nu_{f2}+\nu_{f1}} + \left(M_s\Sigma_{j=0}^{\nu_{f2}+\nu_{f1}-1}\Phi_{\Delta_s}^{j}\Gamma_{\Delta_s} + M_{f1}\Sigma_{j=0}^{\nu_{f2}+\nu_{f1}-1}\Phi_{\Delta_{f1}}^{j}\Gamma_{\Delta_{f1}} + M_{f2}\Sigma_{j=0}^{\nu_{f2}+\nu_{f1}-1}\Phi_{\Delta_{f2}}^{j}\Gamma_{\Delta_{f2}}\right)F_2 \\ \vdots \\ M_{f1}\Phi_{\Delta_{f1}}^{N-1} + \left(M_s\Sigma_{j=0}^{N-2}\Phi_{\Delta_s}^{j}\Gamma_{\Delta_s} + M_{f1}\Sigma_{j=0}^{N-2}\Phi_{\Delta_{f1}}^{j}\Gamma_{\Delta_{f1}} + M_{f2}\Sigma_{j=0}^{N-2}\Phi_{\Delta_{f2}}^{j}\Gamma_{\Delta_{f2}}\right)F_2 \end{bmatrix},$$

$$M_{f2_1} = \begin{bmatrix} M_{f2} \\ M_{f2}\Phi_{\Delta_{f2}} + \left(M_s\Gamma_{\Delta_s} + M_{f1}\Gamma_{\Delta_{f1}} + M_{f2}\Gamma_{\Delta_{f2}}\right)F_3 \\ \vdots \\ M_{f2}\Phi_{\Delta_{f2}}^{\nu_{f2}-1} + \left(M_s\Sigma_{j=0}^{\nu_{f2}-2}\Phi_{\Delta_s}^{j}\Gamma_{\Delta_s} + M_{f1}\Sigma_{j=0}^{\nu_{f2}-2}\Phi_{\Delta_{f1}}^{j}\Gamma_{\Delta_{f1}} + M_{f2}\Sigma_{j=0}^{\nu_{f2}-2}\Phi_{\Delta_{f2}}^{j}\Gamma_{\Delta_{f2}}\right)F_3 \end{bmatrix},$$

$$M_{f2_2} = \begin{bmatrix} M_{f2}\Phi_{\Delta_{f2}}^{\nu_{f2}} + \left(M_s\Sigma_{j=0}^{\nu_{f2}-1}\Phi_{\Delta_s}^{j}\Gamma_{\Delta_s} + M_{f1}\Sigma_{j=0}^{\nu_{f2}-1}\Phi_{\Delta_{f1}}^{j}\Gamma_{\Delta_{f1}} + M_{f2}\Sigma_{j=0}^{\nu_{f2}-1}\Phi_{\Delta_{f2}}^{j}\Gamma_{\Delta_{f2}}\right)F_3 \\ \vdots \\ M_{f2}\Phi_{\Delta_{f2}}^{\nu_{f2}+\nu_{f1}-1} + \left(M_s\Sigma_{j=0}^{\nu_{f2}+\nu_{f1}-2}\Phi_{\Delta_s}^{j}\Gamma_{\Delta_s} + M_{f1}\Sigma_{j=0}^{\nu_{f2}+\nu_{f1}-2}\Phi_{\Delta_{f1}}^{j}\Gamma_{\Delta_{f1}} + M_{f2}\Sigma_{j=0}^{\nu_{f2}+\nu_{f1}-2}\Phi_{\Delta_{f2}}^{j}\Gamma_{\Delta_{f2}}\right)F_3 \end{bmatrix},$$

$$M_{f2_3} = \begin{bmatrix} M_{f2}\Phi_{\Delta_{f2}}^{\nu_{f2}+\nu_{f1}} + \left(M_s\Sigma_{j=0}^{\nu_{f2}+\nu_{f1}-1}\Phi_{\Delta_s}^{j}\Gamma_{\Delta_s} + M_{f1}\Sigma_{j=0}^{\nu_{f2}+\nu_{f1}-1}\Phi_{\Delta_{f1}}^{j}\Gamma_{\Delta_{f1}} + M_{f2}\Sigma_{j=0}^{\nu_{f2}+\nu_{f1}-1}\Phi_{\Delta_{f2}}^{j}\Gamma_{\Delta_{f2}}\right)F_3 \\ \vdots \\ M_{f2}\Phi_{\Delta_{f2}}^{N-1} + \left(M_s\Sigma_{j=0}^{N-2}\Phi_{\Delta_s}^{j}\Gamma_{\Delta_s} + M_{f1}\Sigma_{j=0}^{N-2}\Phi_{\Delta_{f1}}^{j}\Gamma_{\Delta_{f1}} + M_{f2}\Sigma_{j=0}^{\nu_{f2}+\nu_{f1}-2}\Phi_{\Delta_{f2}}^{j}\Gamma_{\Delta_{f2}}\right)F_3 \end{bmatrix}.$$

With (6.22) and (6.25), (6.24) can be manipulated to obtain the sub–matrices of the FOS gain matrix as given in Table 6.2.

Table 6.2 Sub–matrices of the FOS Gain Matrix (L) in (6.23).

$$
\begin{aligned}
L_s = & \left[\tilde{F}_1 - \tilde{F}_3 M_{f2_1}^{-1} M_{s_1} - \left(\tilde{F}_2 - \tilde{F}_3 M_{f2_1}^{-1} M_{f1_1}\right)\left(M_{f1_2} - M_{f2_2} M_{f2_1}^{-1} M_{f1_1}\right)^{-1}\left(M_{s_2} - M_{f2_2} M_{f2_1}^{-1} M_{s_1}\right)\right] \\
& \times\left[\left(M_{s_3} - M_{f2_3} M_{f2_1}^{-1} M_{s_1}\right) - \left(M_{f1_3} - M_{f2_2} M_{f2_1}^{-1} M_{f1_1}\right)\left(M_{f1_2} - M_{f2_2} M_{f2_1}^{-1} M_{f1_1}\right)^{-1}\left(M_{s_2} - M_{f2_2} M_{f2_1}^{-1} M_{s_1}\right)\right]^{-1}, \\
L_{f1} = & \left[\tilde{F}_2 - \tilde{F}_3 M_{f2_1}^{-1} M_{f1_1} - L_s\left(M_{f1_3} - M_{f2_3} M_{f2_1}^{-1} M_{f1_1}\right)\right]\left(M_{f1_2} - M_{f2_2} M_{f2_1}^{-1} M_{f1_1}\right)^{-1}, \\
L_{f2} = & \left(\tilde{F}_3 - L_s M_{f2_3} - L_{f1} M_{f2_2}\right) M_{f2_1}^{-1}.
\end{aligned}
$$

Since F stabilizes (Φ_τ, Γ_τ) and L is an exact solution of (6.9), the closed loop system will also be stable.

6.3 Application to the AHWR

The discrete time equivalent model of (3.3) can be obtained as

$$z_{k+1} = \Phi z_k + \Gamma_1 u_k + \Gamma_2 \delta q_{f_k}, \tag{6.29}$$

$$y_k = M z_k; \tag{6.30}$$

where $\Phi = e^{(A - B_1 K_G M_G)T}$, $\Gamma_1 = \int_{\tau=0}^{T} e^{(A - B_1 K_G M_G)\tau} B_1 d\tau$ and $\Gamma_2 = \int_{\tau=0}^{T} e^{(A - B_1 K_G M_G)\tau} B_2 d\tau$ where T is the sampling interval. It may be reckoned that the discrete model given by (6.29) also exhibits three–time–scale property for sampling interval $T \geq 10s$. Hence the $\triangle$ interval is chosen as $10s$, and the corresponding discrete time model (6.29) is block diagonalized by following the procedure described in Sec. 5.2 into a slow subsystem of order 38, fast 1 subsystem of order 35 and a fast 2 subsystem of order 17, with the state vector partitioned as

$$
\begin{aligned}
z_1 &= \left[z_H^T \; z_X^T \; z_I^T\right]^T, \\
z_2 &= \left[\frac{\delta h_d}{h_{d0}} \; z_C^T \; z_x^T\right]^T, \\
z_3 &= \left[z_Q^T\right]^T;
\end{aligned}
$$

and the Φ, Γ and M matrices partitioned accordingly. The eigenvalues of the original system and the decoupled subsystems are as per Table 5.1, which shows very good agreement between them [96].

The observability index of the AHWR model is 6, with those for slow, fast 1 and fast 2 subsystems respectively being 3, 2 and 1. This prompted the selection of $N = 6$ and the sampling intervals as $\triangle = 10s$ and $\tau = 60s$. It may be noted that in this application the FOS controller is intended to suppress spatial instabilities which have time periods of the order of hours, and a total power controller given by (3.2) is being applied at every fine interval of time on a semi continuous basis for controlling the fast transients in the total power.

Likewise, the τ system is also decoupled into three subsystems, with eigenvalues as listed in Table 6.3. A state feedback gain F is designed for the τ system as described in Sec. 5.4.2, so as to place the closed loop eigenvalues at the locations listed in Table 6.4. For these locations of eigenvalues, system is stable as well as the closed loop system maintains three–time–scale structure. With these subsystem state feedback gains, the FOS gain was designed for the three–time–scale model of

Table 6.3 Comparison of Original and Subsystem Eigenvalues of the AHWR Model with Total Power Feedback, Discretized with 60s sampling interval.

k	ψ_k(orig. system)	ψ_k(subsystem)	k	ψ_k(orig. system)	ψ_k(subsystem)
		Slow Subsystem	50	2.4318×10^{-2}	2.4318×10^{-2}
$1-2$	$1.0053\pm j1.3149\times10^{-3}$	$1.0053\pm j1.3149\times10^{-3}$	51	2.3945×10^{-2}	2.3945×10^{-2}
$3-4$	$1.0048\pm j2.4034\times10^{-3}$	$1.0048\pm j2.4034\times10^{-3}$	52	2.3688×10^{-2}	2.3688×10^{-2}
5	9.9945×10^{-1}	9.9945×10^{-1}	53	2.3578×10^{-2}	2.3578×10^{-2}
6	9.9938×10^{-1}	9.9938×10^{-1}	54	2.3366×10^{-2}	2.3366×10^{-2}
7	9.9912×10^{-1}	9.9912×10^{-1}	55	2.3008×10^{-2}	2.3008×10^{-2}
8	9.9847×10^{-1}	9.9847×10^{-1}	56	2.2969×10^{-2}	2.2969×10^{-2}
$9-10$	$9.9782\times10^{-1}\pm j4.5870\times10^{-3}$	$9.9782\times10^{-1}+4.5870\times10^{-3}$	57	8.8611×10^{-4}	8.8611×10^{-4}
11	9.9774×10^{-1}	9.9774×10^{-1}	58	1.4672×10^{-4}	1.4672×10^{-4}
$12-13$	$9.9772\times10^{-1}\pm j4.5781\times10^{-3}$	$9.9772\times10^{-1}+4.5781\times10^{-3}$	59	1.4657×10^{-4}	1.4657×10^{-4}
14	9.9772×10^{-1}	9.9772×10^{-1}	60	1.3842×10^{-4}	1.3842×10^{-4}
15	9.9760×10^{-1}	9.9760×10^{-1}	61	1.3507×10^{-4}	1.3507×10^{-4}
16	9.9751×10^{-1}	9.9751×10^{-1}	62	8.7139×10^{-5}	8.7139×10^{-5}
17	9.9747×10^{-1}	9.9747×10^{-1}	63	8.6891×10^{-5}	8.6891×10^{-5}
18	9.9735×10^{-1}	9.9735×10^{-1}	64	8.2945×10^{-5}	8.2945×10^{-5}
19	9.9719×10^{-1}	9.9719×10^{-1}	65	7.8213×10^{-5}	7.8213×10^{-5}
20	9.9707×10^{-1}	9.9707×10^{-1}	66	5.6017×10^{-5}	5.6017×10^{-5}
$21-22$	$9.9611\times10^{-1}\pm j3.1742\times10^{-3}$	$9.9611\times10^{-1}\pm j3.1742\times10^{-3}$	67	5.5746×10^{-5}	5.5746×10^{-5}
$23-24$	$9.9604\times10^{-1}\pm j3.2847\times10^{-3}$	$9.9604\times10^{-1}\pm j3.2847\times10^{-3}$	68	5.3113×10^{-5}	5.3113×10^{-5}
$25-26$	$9.9561\times10^{-1}\pm j2.3488\times10^{-3}$	$9.9561\times10^{-1}\pm j2.3488\times10^{-3}$	69	4.7879×10^{-5}	4.7879×10^{-5}
$27-28$	$9.9536\times10^{-1}\pm j1.7874\times10^{-3}$	$9.9536\times10^{-1}\pm j1.7874\times10^{-3}$	70	1.9949×10^{-5}	1.9949×10^{-5}
29	9.9157×10^{-1}	9.9157×10^{-1}	71	1.9806×10^{-5}	1.9806×10^{-5}
30	9.9128×10^{-1}	9.9128×10^{-1}	72	1.8963×10^{-5}	1.8963×10^{-5}
31	9.9061×10^{-1}	9.9061×10^{-1}	73	1.6058×10^{-5}	1.6058×10^{-5}
32	9.9013×10^{-1}	9.9013×10^{-1}			Fast 2 Subsystem
33	9.9006×10^{-1}	9.9006×10^{-1}	74	0.0	0.0
34	9.8967×10^{-1}	9.8967×10^{-1}	75	0.0	0.0
35	9.8878×10^{-1}	9.8878×10^{-1}	76	0.0	0.0
36	9.8874×10^{-1}	9.8874×10^{-1}	77	0.0	0.0
$37-38$	9.6677×10^{-1}	9.6641×10^{-1}	78	0.0	0.0
		Fast 1 Subsystem	79	0.0	0.0
39	9.9684×10^{-2}	9.9028×10^{-2}	80	0.0	0.0
40	8.8464×10^{-2}	8.8464×10^{-2}	81	0.0	0.0
41	4.7016×10^{-2}	4.7016×10^{-2}	82	0.0	0.0
42	4.6441×10^{-2}	4.6441×10^{-2}	83	0.0	0.0
43	3.1309×10^{-2}	3.1309×10^{-2}	84	0.0	0.0
44	3.1007×10^{-2}	3.1007×10^{-2}	85	0.0	0.0
45	2.7808×10^{-2}	2.7808×10^{-2}	86	0.0	0.0
46	2.7781×10^{-2}	2.7781×10^{-2}	87	0.0	0.0
47	2.6766×10^{-2}	2.6766×10^{-2}	88	0.0	0.0
48	2.6291×10^{-2}	2.6291×10^{-2}	89	0.0	0.0
49	2.4456×10^{-2}	2.4456×10^{-2}	90	0.0	0.0

Table 6.4 Comparison of Closed Loop Eigenvalues of the Original and Subsystems.

k	$\psi_k(\Phi+\Gamma_1 F)$	Subsystems	k	$\psi_k(\Phi+\Gamma_1 F)$	Subsystems
		$\psi_k\left(\Phi_{\tau_s}+\overline{\Gamma}_{\tau_s}F_s\right)$	50	2.3688×10^{-2}	2.3688×10^{-2}
1	9.9995×10^{-1}	9.9992×10^{-1}	51	2.3577×10^{-2}	2.3578×10^{-2}
$2-3$	$9.9787\times 10^{-1}\pm j4.6369\times 10^{-3}$	$9.9787\times 10^{-1}\pm j4.6365\times 10^{-3}$	52	2.3366×10^{-2}	2.3366×10^{-2}
4	9.9774×10^{-1}	9.9774×10^{-1}	53	2.3008×10^{-2}	2.3008×10^{-2}
5	9.9772×10^{-1}	9.9772×10^{-1}	54	2.2969×10^{-2}	2.2970×10^{-2}
$6-7$	$9.9772\times 10^{-1}\pm j4.5747\times 10^{-3}$	$9.9772\times 10^{-1}\pm j4.5747\times 10^{-3}$	55	2.2100×10^{-2}	2.2109×10^{-2}
8	9.9760×10^{-1}	9.9760×10^{-1}	56	7.6710×10^{-3}	7.8716×10^{-3}
9	9.9751×10^{-1}	9.9751×10^{-1}	57	8.8549×10^{-4}	8.8311×10^{-4}
10	9.9747×10^{-1}	9.9747×10^{-1}	58	1.4672×10^{-4}	1.4671×10^{-4}
11	9.9735×10^{-1}	9.9735×10^{-1}	59	1.4656×10^{-4}	1.4656×10^{-4}
12	9.9716×10^{-1}	9.9716×10^{-1}	60	1.3841×10^{-4}	1.3841×10^{-4}
13	9.9707×10^{-1}	9.9707×10^{-1}	61	1.3498×10^{-4}	1.3354×10^{-4}
$14-15$	$9.9611\times 10^{-1}\pm j3.1712\times 10^{-3}$	$9.9611\times 10^{-1}\pm j3.1712\times 10^{-3}$	62	8.7138×10^{-5}	8.7134×10^{-5}
$16-17$	$9.9605\times 10^{-1}\pm j3.2709\times 10^{-3}$	$9.9605\times 10^{-1}\pm j3.2710\times 10^{-3}$	63	8.6889×10^{-5}	8.6889×10^{-5}
$18-19$	$9.9560\times 10^{-1}\pm j2.3476\times 10^{-3}$	$9.9560\times 10^{-1}\pm j2.3476\times 10^{-3}$	64	8.2938×10^{-5}	8.2938×10^{-5}
$20-21$	$9.9536\times 10^{-1}\pm j1.7866\times 10^{-3}$	$9.9536\times 10^{-1}\pm j1.7866\times 10^{-3}$	65	7.7855×10^{-5}	7.7367×10^{-5}
$22-23$	$9.9517\times 10^{-1}\pm j2.3530\times 10^{-3}$	$9.9517\times 10^{-1}\pm j2.3530\times 10^{-3}$	66	5.6016×10^{-5}	5.6016×10^{-5}
$24-25$	$9.9470\times 10^{-1}\pm j1.2415\times 10^{-3}$	$9.9470\times 10^{-1}\pm j1.2415\times 10^{-3}$	67	5.5745×10^{-5}	5.5745×10^{-5}
26	9.9157×10^{-1}	9.9157×10^{-1}	68	5.3107×10^{-5}	5.3107×10^{-5}
27	9.9127×10^{-1}	9.9127×10^{-1}	69	4.7275×10^{-5}	4.6259×10^{-5}
28	9.9061×10^{-1}	9.9061×10^{-1}	70	1.9949×10^{-5}	1.9948×10^{-5}
29	9.9013×10^{-1}	9.9013×10^{-1}	71	1.9805×10^{-5}	1.9805×10^{-5}
30	9.9002×10^{-1}	9.9002×10^{-1}	72	1.8961×10^{-5}	1.8961×10^{-5}
31	9.8967×10^{-1}	9.8967×10^{-1}	73	1.5354×10^{-5}	1.5267×10^{-5}
32	9.8878×10^{-1}	9.8878×10^{-1}			$\psi_{k-73}\left(\Phi_{\tau_{f2}}+\Gamma_{\tau_{f2}}F_{f2}\right)$
33	9.8874×10^{-1}	9.8874×10^{-1}	74	0.00	0.00
34	6.7206×10^{-1}	6.7208×10^{-1}	75	0.00	0.00
35	6.3534×10^{-1}	6.3587×10^{-1}	76	0.00	0.00
36	6.1260×10^{-1}	6.1536×10^{-1}	77	0.00	0.00
37	5.5020×10^{-1}	5.5110×10^{-1}	78	0.00	0.00
38	5.4710×10^{-1}	5.4820×10^{-1}	79	0.00	0.00
		$\psi_{k-38}\left(\Phi_{\tau_{f1}}+\overline{\Gamma}_{\tau_{f1}}F_{f1}\right)$	80	0.00	0.00
39	4.6965×10^{-2}	4.6965×10^{-2}	81	0.00	0.00
40	4.6399×10^{-2}	4.6399×10^{-2}	82	0.00	0.00
41	3.1310×10^{-2}	3.1301×10^{-2}	83	0.00	0.00
42	3.1003×10^{-2}	3.1003×10^{-2}	84	0.00	0.00
43	2.7808×10^{-2}	2.7809×10^{-2}	85	0.00	0.00
44	2.7780×10^{-2}	2.7780×10^{-2}	86	0.00	0.00
45	2.6765×10^{-2}	2.6765×10^{-2}	87	0.00	0.00
46	2.6291×10^{-2}	2.6293×10^{-2}	88	0.00	0.00
47	2.4456×10^{-2}	2.4456×10^{-2}	89	0.00	0.00
48	2.4319×10^{-2}	2.4316×10^{-2}	90	0.00	0.00
49	2.3945×10^{-2}	2.3945×10^{-2}			

AHWR by decomposing the problem into three subsystem problems as described in the previous section.

Table 6.5 lists the FOS gain matrices L_{f2}, L_{f1} and L_s, of respective dimensions 4×17, 4×51 and 4×34, computed using these subsystem gains and the subsystem matrices of the Δ–system applying the described method. It can be verified that with these gains, the closed loop eigenvalues of the full order τ–system are exactly equal to those with state feedback. It may further be noticed that the gains corresponding to fastest subsystem are all zeros, and the maximum values of the gains for the fast 1 and slow subsystems are only 3.1490 and 3.2430 respectively, which seem to be acceptable for practical implementation.

Table 6.5 Subsystem FOS Gain Matrices.

$$L_{f2} = \begin{bmatrix} 0 \dots 0 \\ 0 \dots 0 \\ 0 \dots 0 \\ 0 \dots 0 \end{bmatrix},$$

$$L_{f1} = \left[\begin{array}{rrrrrrrrrrrr} -0.1506 & 0.0544 & 0.0668 & 0.0631 & 0.0322 & -0.0020 & 0.0234 & -0.0669 & 0.0353 & -0.0913 & 0.0509 & 0.0397 \\ -0.0629 & 0.0176 & 0.0526 & 0.0473 & 0.0364 & -0.0189 & 0.0051 & -0.0821 & -0.0081 & -0.0189 & 0.0235 & -0.0293 \\ 0.1669 & -0.0647 & -0.0695 & -0.0373 & -0.0424 & -0.0036 & -0.0112 & 0.0650 & -0.0374 & 0.0973 & -0.0474 & -0.0437 \\ 0.1223 & -0.0340 & -0.0641 & -0.0416 & -0.0622 & 0.0208 & -0.0172 & 0.0956 & 0.0045 & 0.0389 & -0.0396 & 0.0266 \end{array} \right.$$

$$\begin{array}{rrrrrrrrrrrr} 0.0459 & -0.0568 & 0.1437 & -0.1030 & -0.0642 & 1.216 & -0.4569 & -0.4201 & -0.4308 & -0.5424 & -0.0315 & -0.2211 \\ -0.0272 & -0.0531 & 0.1002 & 0.0281 & 0.0201 & 0.5589 & -0.2547 & -0.3204 & -0.2633 & -0.3382 & 0.2342 & -0.0794 \\ -0.0228 & 0.0511 & -0.1348 & 0.0876 & 0.0668 & -1.1280 & 0.4578 & 0.4035 & 0.2631 & 0.5109 & -0.0038 & 0.1376 \\ 0.0439 & 0.0708 & -0.1267 & -0.0229 & -0.0018 & -0.7862 & 0.3246 & 0.3641 & 0.2115 & 0.4820 & -0.3308 & 0.1876 \end{array}$$

$$\begin{array}{rrrrrrrrrrrr} 0.5387 & -0.0715 & 0.7108 & -0.5390 & -0.3966 & -0.2281 & 0.5735 & -1.2401 & 0.9295 & 0.4278 & -3.1490 & 1.0961 \\ 0.5923 & 0.0304 & 0.2032 & -0.3080 & 0.0981 & 0.1618 & 0.4418 & -0.8151 & -0.0956 & -0.0635 & -1.5210 & 0.8101 \\ -0.5244 & 0.0262 & -0.7008 & 0.4883 & 0.4091 & 0.1408 & -0.5164 & 1.1370 & -0.7936 & -0.4052 & 2.7360 & -0.9599 \\ -0.7325 & -0.0784 & -0.3142 & 0.4331 & -0.0739 & -0.2109 & -0.5911 & 1.0350 & 0.0602 & -0.0235 & 1.9040 & -1.0070 \end{array}$$

$$\begin{array}{rrrrrrrrrrrr} 0.5353 & 1.0090 & 1.8611 & 0.1961 & 0.9376 & -1.5210 & -0.1636 & -1.6930 & 1.7771 & 1.0822 & 0.3120 & -1.8300 \\ 0.4089 & -0.0019 & 0.9410 & -0.8480 & 0.5004 & -0.9021 & -0.0457 & -0.6339 & 1.2941 & 0.5036 & -0.1201 & -1.1220 \\ -0.4870 & -0.7316 & -1.6720 & 0.0396 & -0.8132 & 1.3511 & 0.3501 & 1.5320 & -1.7001 & -1.1381 & -0.1640 & 1.5862 \\ -0.5026 & 0.1690 & -1.2700 & 1.1221 & -0.8559 & 1.2970 & 0.2883 & 0.8457 & -1.6740 & -0.6736 & 0.1821 & 1.4720 \end{array}$$

$$\left. \begin{array}{rrr} 3.0170 & -2.5071 & -0.7935 \\ 1.7250 & -0.5518 & -0.1813 \\ -2.7530 & 2.1401 & 0.7023 \\ -2.3330 & 0.5853 & 0.3368 \end{array} \right],$$

$$L_s = \left[\begin{array}{rrrrrrrrrrrr} 3.2430 & -0.9765 & 0.2642 & -1.0240 & -2.3161 & -0.4072 & -1.6380 & 1.8071 & 0.4975 & 1.5070 & -2.3670 & -1.1171 \\ 1.6001 & -0.9586 & 0.1859 & 1.0760 & -0.9913 & 1.1601 & -1.0290 & -0.1572 & 0.0195 & 0.7573 & -2.0610 & -1.6650 \\ -2.7670 & 0.6857 & -0.2924 & 0.9358 & 2.0820 & 0.7438 & 1.5990 & -1.4360 & -0.7192 & -1.2001 & 2.3640 & 1.1910 \\ -1.9540 & 1.2651 & -0.0743 & -1.3200 & 1.3271 & -1.3670 & 1.4680 & -0.3079 & -0.3180 & -0.9270 & 2.5150 & 1.9741 \end{array} \right.$$

$$\begin{array}{rrrrrrrrrrrr} -0.0186 & 2.3150 & -2.6560 & 2.6021 & 0.4724 & -1.1590 & 0.2788 & -0.4493 & 0.3821 & 0.9671 & 0.2476 & 0.9001 \\ -0.3022 & 1.1080 & -1.0970 & 1.7881 & 0.6560 & -0.5764 & 0.3853 & -0.3307 & -0.8636 & 0.3481 & -0.5277 & 0.6059 \\ -0.0915 & -1.9380 & 2.4121 & -2.2330 & -0.3516 & 0.9916 & -0.1158 & 0.4476 & -0.4305 & -0.8808 & -0.0434 & -0.9156 \\ 0.2767 & -1.4300 & 1.7670 & -1.7391 & -0.7626 & 0.7136 & -0.5500 & 0.2797 & 0.9856 & -0.4745 & 0.5552 & -0.7860 \end{array}$$

$$\left. \begin{array}{rrrrrrrrrr} -0.7590 & -0.3004 & -0.4375 & 1.0740 & 0.3927 & -0.1091 & -0.9990 & 0.7371 & -0.9224 & -0.0456 \\ 0.5535 & 0.0071 & -0.3071 & 1.0490 & 1.0880 & 0.2842 & -0.3766 & 0.0892 & -1.1650 & -0.4287 \\ 0.5440 & 0.3822 & 0.2747 & -1.1020 & -0.4183 & 0.1344 & 0.8131 & -0.6648 & 0.7991 & -0.0101 \\ -0.3571 & 0.1036 & 0.3552 & -1.2320 & -1.2490 & -0.2892 & 0.4786 & -0.3460 & 1.1120 & 0.4477 \end{array} \right]$$

6.4 Transient Response

The efficacy of the controller was evaluated through simulations. Total power control as per the control law (3.2) was applied every $5ms$. For spatial power control, output power levels were sampled every $\Delta = 10s$ to formulate the Y matrix given by (6.3). New values of spatial power control signals were generated every $\tau = 60s$ and applied to the RR drives alongwith the signal generated by the total power controller. Fig. 6.1 depicts the control signals generated by FOS during a spatial power transient identical to that described in Sec. 3.4. The corresponding RR positions are shown in Fig. 6.2. This resulted in perturbations in spatial power distribution, which were suppressed by the controller within about $500s$, as shown in Fig. 6.3. Fig. 6.4 depicts the variations in quadrant core power distribution during the transient. The total reactor power underwent an initial overshoot by about 0.5% which was immediately stabilized, as depicted in Fig. 6.5. The response of the controller is found to

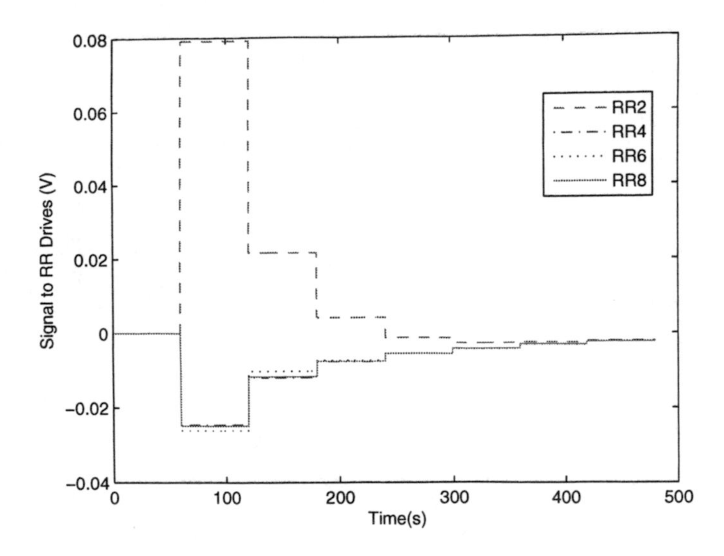

Fig. 6.1 Control Signals Generated by FOS Controller.

be slow as compared to the state feedback controller obtained in the previous chapter. However, it should be reckoned that here the control output is generated once in 60 s, and even with this, the controller could suppress the disturbances within about 6 samples.

Fig. 6.6 shows the closed loop system response during another transient initiated by a momentary disturbance in position of RR2. It was assumed that the reactor was initially operating at full power, with only the total power controller according to (3.1) active. The RR2, which was initially at its equilibrium position, was driven out by about 1.5% under a control signal of $-1V$, and immediately driven in so as to come back to its original position under a signal of $+1V$. As a consequence of this disturbance, the flux tilts started picking up as shown in Fig. 6.6. After about 15 hours, the spatial control component as per the control law (6.3) was introduced in the control signals. Fig. 6.6 also shows a blown–up plot of the response during 10 minutes immediately after the introduction of the spatial control component, from which it can be observed that the tilts are suppressed within about 10 minutes. The tilts remain suppressed thereafter during the remaining prolonged simulation.

During a transient involving a step disturbance in the feed water flow rate as detailed in Sec. 3.4.1, the response of the closed loop system is found to be stable. The total reactor power stabilized at about 0.14% higher than the original power, as shown in Fig. 6.7. Corresponding variations in nodal powers and exit qualities are depicted respectively in Fig. 6.8 and Fig. 6.9.

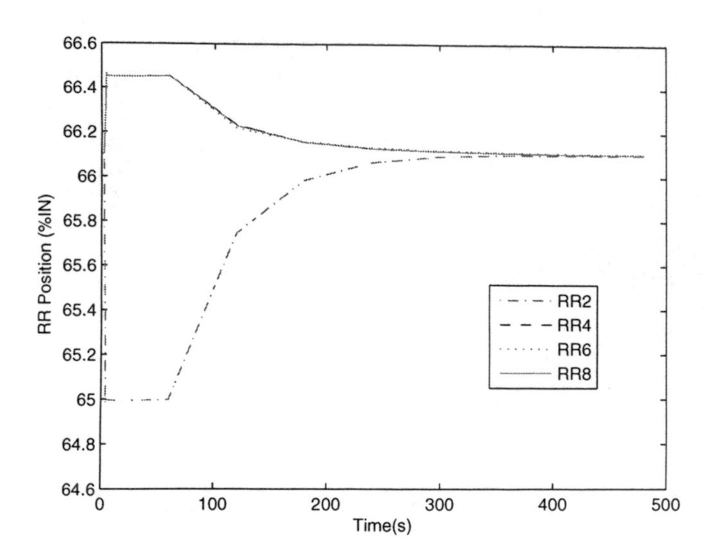

Fig. 6.2 RR Movement with FOS Controller.

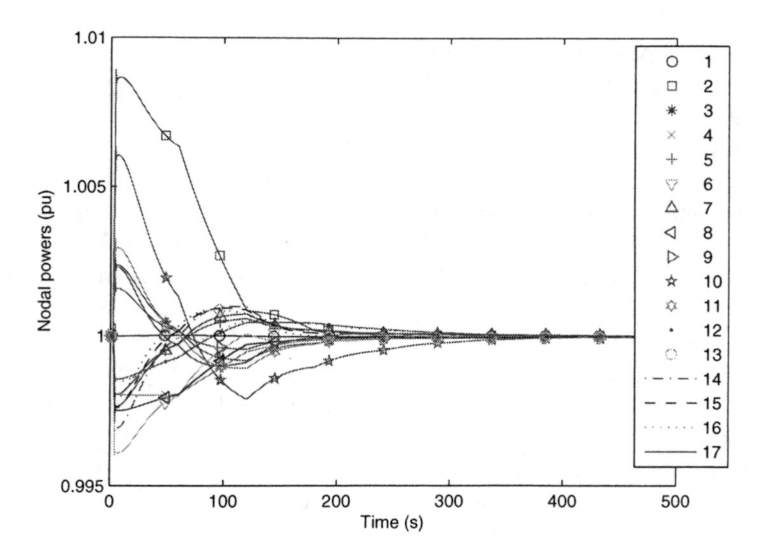

Fig. 6.3 Nodal Power Variations with FOS Controller.

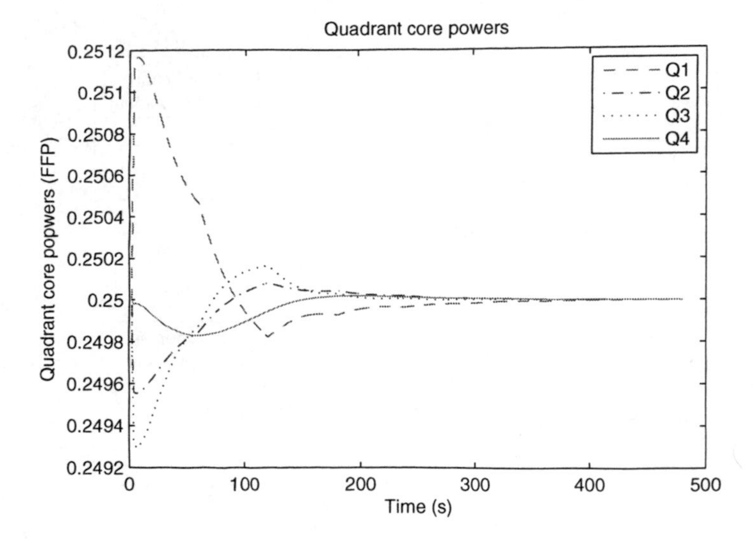

Fig. 6.4 Quadrant Core Power Variations with FOS Controller.

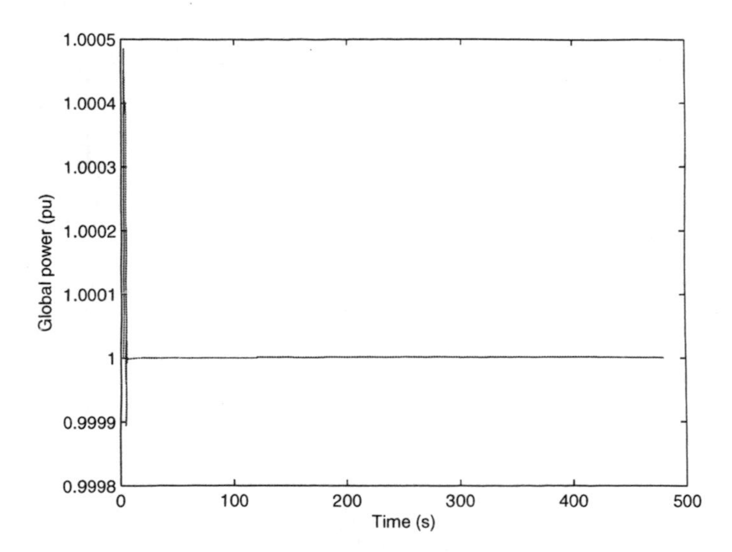

Fig. 6.5 Total Power Variation during the Transient.

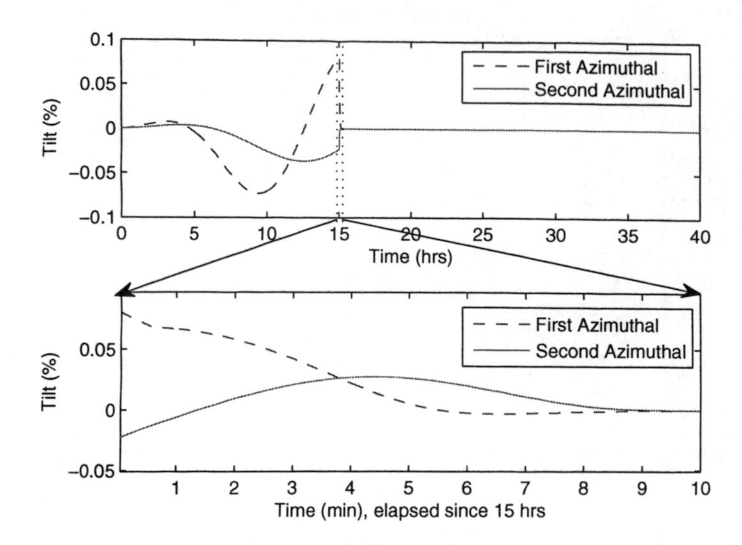

Fig. 6.6 Suppression of Tilts with FOS Controller.

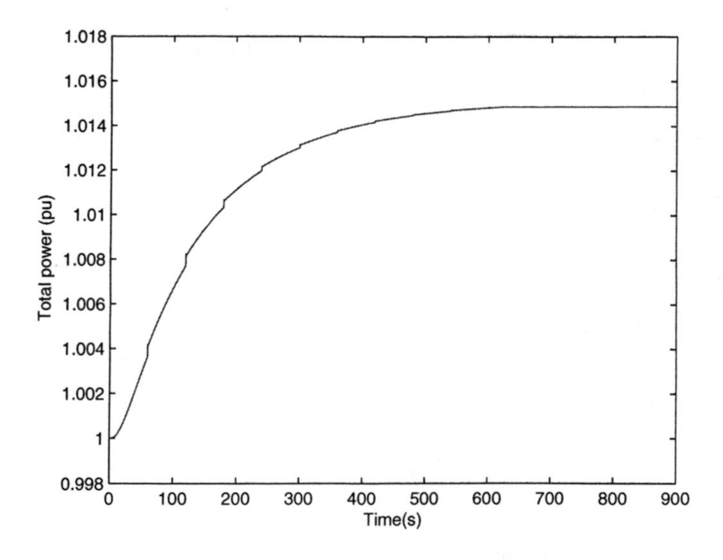

Fig. 6.7 Total Power Variation during Step Change in Feed Flow.

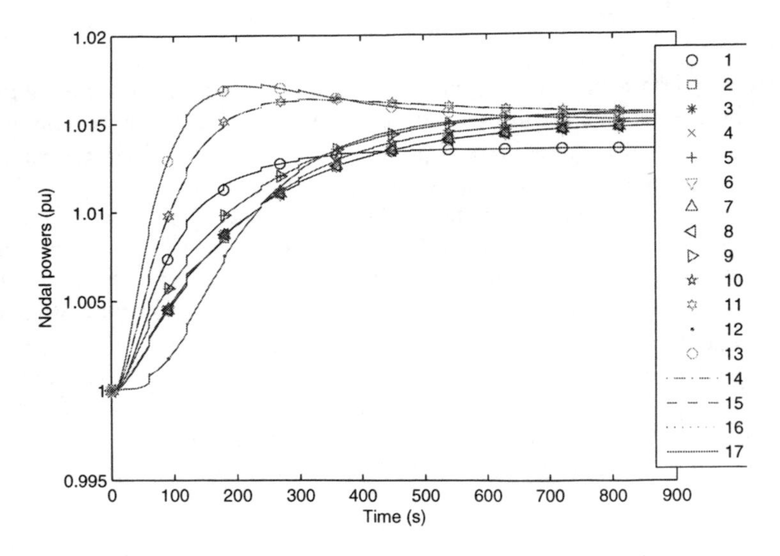

Fig. 6.8 Nodal Power Variations during Step Change in Feed Flow.

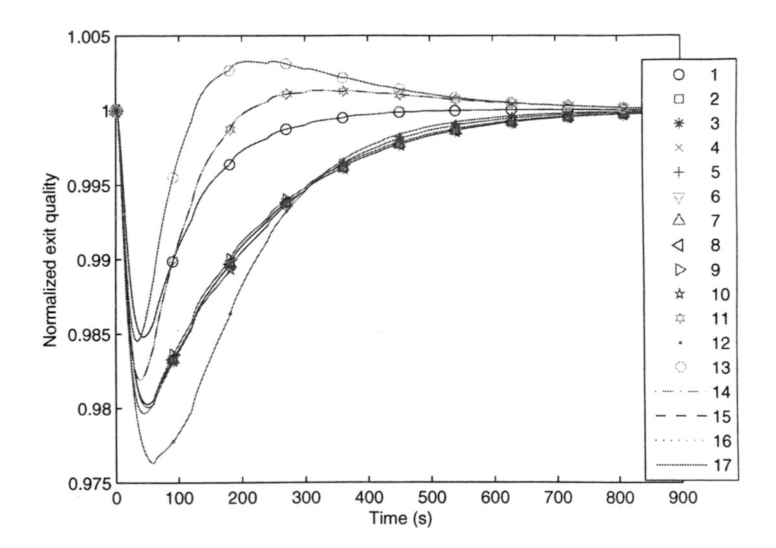

Fig. 6.9 Exit Quality Variations during Step Change in Feed Flow.

6.5 Conclusion

In this chapter a method is presented to compute the fast output sampling gains of higher order systems from the lower order subsystems obtained by decomposing the original higher order system into three subsystems. Thus the ill–conditioning effect of the higher order matrices are avoided in gain computations.

The proposed technique is applied to the spatial control of AHWR, where the system matrix involved is of order 90. It was experienced that the direct computation of the FOS gain matrix was not successful for any sampling intervals, essentially owing to the heavy ill–conditioning. However, the eigenvalues of the original non–linear model of AHWR exhibits a three–time–scale structure, which was retained and exploited by a proper selection of the sampling interval. Subsequently the system is decoupled into three smaller order subsystems with which the FOS gain could be computed. The controller thus designed employs the feedback of only the output information, thereby evading the need of a state observer.

References

1. Aleite, W.: The Power Distribution Controls of BIBLIS–A Nuclear Power Station and Future Developments with Process Computers. In: IAEA Specialists' Meeting of the Working Group on Nucl. Power Plant Contr. and Inst.– Spatial Control Problems, Studsvik, Sweden, October 28-29 (1974)
2. Antoulas, A.C., Sorensen, D.C., Gugercin, S.: A survey of model reduction methods for large scale systems. Contemporary Mathematics 280, 193–219 (2001)
3. Aoki, M.: Control of Large Scale Dynamic Systems by Aggregation. IEEE Trans. Aut. Contr. AC-13(3), 246–253 (1968)
4. Ardema, M.: Singular Perturbations in Systems and Control. CISM Courses and Lectures, vol. 280. Springer, New York (1983)
5. Asatani, K., Iwazumi, T., Hattori, Y.: Error Estimation of Prompt Jump Approximation by Singular Perturbation Theory. Journal of Nucl. Sci. and Tech. 8(11), 653–656 (1971)
6. ASTM E2006-05 Standard Guide for Benchmark Testing of Light Water Reactor Calculations. ASTM Book of Standards, vol. 12.02 (2005)
7. Åström, K.J., Bell, R.D.: Drum-Boiler Dynamics. Automatica 36, 363–378 (2000)
8. Avery, R.: Theory of Coupled Reactors. In: Proc. 2nd Intern. Conf. Peaceful Uses At. Energy, vol. 12, p. 182 (1958)
9. Bandyopadhyay, B., Manjunath, T.C.: Fault tolerant control of flexible smart structures using robust decentralized fast output sampling feedback technique. Asian Jnl. of Ctrl. 9, 268–291 (2007)
10. Bandyopadhyay, B., Sharma, G.L.: Fault-Tolerant Spatial Control of a Large Pressurised Heavy Water Reactor by Fast Output Sampling Technique. Proceedings of IEEE Ctrl. Th. and Appl. 151(1), 117–124 (2004)
11. Belchior Jr., A., Moreira, J.M.L.: Chebyshev Acceleration for Lambda Mode Calculations. Trans. American Nucl. Soc. 66, 237–238 (1992)
12. Benchmark Problem Book. ANL National Energy Software Centre, ANL-7416, pp: 674–793 (1985)
13. Camiciola, P., Cundari, D., Montagnini, B.: A Coarse Mesh Method for 1-D Reactor Kinetics. Annals of Nuclear Energy 13, 629–636 (1986)
14. Canosa, J., Brooks, H.: Xenon–Induced Oscillations. Nucl. Sci. and Engg. 26, 237–253 (1966)
15. Chang, K.W.: Diagonalization Method for a Vector Boundary Problem of Singular Perturbation Type. J. Math. Anal. Appl. 48, 16 (1974)
16. Chen, C.-T.: Linear System Theory and Design, 3rd edn. Oxford University Press, New York (1999)
17. Cho, N.Z., Grossman, L.M.: Optimal Control for Xenon Spatial Oscillations in Load Follow of a Nuclear Reactor. Nucl. Sci. and Engg. 83, 136–148 (1983)
18. Chow, J.H., Kokotovic, P.V.: A Decomposition of Near Optimum Regulators for Systems with Slow and Fast Modes. IEEE Trans. Aut. Ctrl. AC-21, 701 (1976)
19. Computational Benchmark Problems Committee of the Mathematics and Computation Division of the American Nuclear Society, National Energy Software Centre: Benchmark Problem Book, Illinois: ANL–7416, Argonne National Laboratory (December 1985)
20. Difilippo, F.C., Waldman, R.M.: Kinetics of a Coupled Two-Core Nuclear Reactor. Nucl. Sci. and Engg. 61, 60 (1976)
21. DBR on RRS of AHWR. DBR/237/23711/0, BARC (May 2002)
22. Dodds Jr., H.L.: Accuracy of the Quasistatic Method for Two–Dimensional Thermal Reactor Transients with Feedback. Nucl. Sci. and Engg. 59, 271–276 (1976)

23. Dontchev, A.L.: Perturbations, Approximations and Sensitivity Analysis of Optimal Control Systems. LNCIS, vol. 52. Springer, Berlin (1983)
24. Duderstadt, J.J., Hamilton, L.J.: Nuclear Reactor Analysis. John Wiley and Sons, Inc., New York (1976)
25. Eckhaus, W., de Jarger, E.M.: Theory and Applications of Singular Perturbations. Springer, Berlin (1982)
26. Erdelyi, A.: Singular perturbations of boundary value problems involving ordinary differential equations. J. Soc. Ind. Appl. Math. 11, 105 (1963)
27. Farawila, Y.M.: Application of Modal Neutron Kinetics to BWR Oscillation Problems. Nucl. Sci. and Engg. 129(3), 261–272 (1998)
28. Fortuna, L., Nunnari, G., Gallo, A.: Model Order Reduction Techniques in Electrical Engineering. Springer, London (1992)
29. Gaitsgory, V., Nguyen, M.T.: Averaging of Three Time Scale Singularly Perturbed Control Systems. Systems & Control Letters 42 (2001)
30. Garland, W.J., Hand, B.J.: Simplified Functions for the Fast Approximation of Light Water Thermodynamic Properties. Nuclear Engineering and Design 113, 21–34 (1989)
31. Garland, W.J., Wilson, R.J., Bartak, J., Cizek, J., Stasny, M., Zentrich, I.: Extensions to the approximation functions for the fast calculation of saturated water properties. Nuclear Engineering and Design 136, 381–388 (1992)
32. Glasstone, S., Sesonske, A.: Nuclear Reactor Engineering. CBS Publishers & Distributers, Delhi (1986)
33. Glasstone, S., Edlund, M.C.: The Elements of Nuclear Reactor Theory. D. Van Nostrand, New York (1952)
34. Grossman, L.M., Hennart, J.P.: Nodal Diffusion Methods for Space–Time Neutron Kinetics. Progress in Nuclear Energy 49, 181–216 (2007)
35. Haddad, A., Kokotovic, P.V.: Note on Singular Perturbation of Linear State Regulators. IEEE Trans. Aut. Ctrl. AC-16, 279–281 (1971)
36. Hagiwara, T., Araki, M.: Design of a state feedback controller based on multirate sampling of plant output. IEEE Trans. Aut. Ctrl. 33, 812–819 (1988)
37. Hautus, M.L.J.: Stabilization Controllability and Observability of Linear Autonomous Systems. Nederl. Akad. Wetensch. Proc. Ser. A72, 448–455 (1970)
38. Hetrick, D.L.: Dynamics of Nuclear Reactors. The University of Chicago Press, Chicago (1971)
39. Hideaki, I., Toshikazu, T.: Development and Verification of an Efficient Spatial Neutron Kinetics Method for Reactivity Initiated Event Analysis. Nucl. Sci. and Tech. 38(7), 492–502 (2001)
40. Hinchley, E., Kuglar, G.: On–line Control of the CANDU–PHW Power Distribution. In: IAEA Specialists' Meeting on Spatial Control Problems, Studsvik, Sweden (October 1974)
41. Huo, X., Xie, Z., Liao, C.: Development of Core Fuel Management Code System for CANDU Reactors based on Coupled Neutronics and Thermal Hydraulics Advanced Nodal Method. Annals Nucl. Energy 31(10), 1083–1180 (2004)
42. Hutton, M.F., Friedland, B.: Routh Approximation for Reducing Order of Linear Time Invariant Systems. IEEE Trans. Aut. Contr. AC-20(3), 329–337 (1975)
43. Jamshidi, M.: Large Scale Systems: Modeling and Control. North-Holland Systems Science and Engineering, Elsevier Science Pub. Co. Inc., New York (1983)
44. Janardhanan, S., Bandyopadhyay, B.: Fast Output Sampling based Output Feedback Sliding Mode Control Law for Uncertain Systems. In: Proc. 3rd Int. Conf. System Identification and Control Problems (SICPRO 2004), Moscow, pp. 966–974 (January 2004)

45. Javid, S.H.: Uniform Asymptotic Stability of Linear Time Varying Singularly Perturbed Systems. J. Franklin Inst. 305, 27 (1978)
46. Judd, R.A., Rouben, B.: Three Dimensional Kinetics Benchmark Problem in Heavy Water Reactors. Report No. AECL-7236 (1981)
47. Kando, H., Iwazumi, T., Ukai, H.: Singular Perturbation Modelling of Large–Scale Systems with Multi–Time–Scale Property. Int. J. Ctrl. 48, 6 (1988)
48. Kang, C.M., Hansen, K.F.: Finite Element Methods for Reactor Analysis. Nucl. Sci. and Engg. 51, 456–495 (1973)
49. Kaplan, S., Marlowe, O.J., Bewick, J.: Application of Synthesis Techniques to Problems Involving Time Dependence. Nucl. Sci. and Engg. 18, 163 (1964)
50. Karppinen, J.: Spatial Reactor Control Methods. Nucl. Sci. and Engg. 64, 657–672 (1977)
51. Kevorkian, J., Cole, J.D.: Multiple scale and singular perturbation methods, 1st edn. Springer (May 1996)
52. Kobayashi, K.: Rigorous Derivation of Static and Kinetic Nodal Equations for Coupled Reactors Using Transport Theorem. Jnl. Nucl. Sci. Tech. 28(5), 389–398 (1991)
53. Kobayashi, K., Yoshikuni, M.: Analysis of Xenon Oscillation by Coupled Reactor Model. Jnl. Nucl. Sci. and Tech. 19(2), 107–118 (1982)
54. Kokotovic, P.V.: Subsystems, Time-Scales and Multi–Modeling. Automatica 19, 789 (1981)
55. Kokotovic, P.V., Allemong, J.J., Winkelman, J.R., Chow, J.H.: Singular Perturbation and Iterative Separation of Time Scales. Automatica 16, 23 (1980)
56. Kokotovic, P.V., O'Malley Jr., R.E., Sannuti, P.: Singular Perturbation and Order Reduction in Control Applications – An Overview. Automatica 12, 123 (1976)
57. Komata, M.: On the Derivation of Avery's Coupled Reactor Kinetics Equations. Nucl. Sci. and Engg. 38, 193–204 (1969)
58. Ladde, G.S., Rajalakshmi, S.G.: Diagonalization and Stability of Multi–Time–Scale Singularly Perturbed Linear Systems. Appl. Math. and Comp. 16, 115 (1985)
59. Ladde, G.S., Siljak, D.D.: Multiparameter Singular Perturbations of Linear Systems with Multiple Time Scales. Automatica 19, 4 (1983)
60. Lamarsh, J.R.: Introduction to Nuclear Reactor Theory. Addison–Wesley, Reading (1966)
61. Langenbuch, S., Maurer, W., Werner, W.: High–Order Schemes for Neutron Kinetics Calculations Based on Local Polynomial Approximation. Nucl. Sci. and Engg. 64, 508–516 (1977)
62. Latkouhi, B., Khalil, H.: Multirate and Composite Control of Two–Time–Scale Discrete–Time Systems. IEEE Trans. Aut. Ctrl. AC-30, 7 (1985)
63. Mahmoud, M.S., Singh, M.G.: Large Scale System Modeling. Pergamon Press, Oxford (1981)
64. Mahmoud, M.S., Chen, Y., Singh, M.G.: On Eigenvalue Assignment in Discrete Systems with Fast and Slow Modes. Int. J. Systems Sci. 16, 1 (1985)
65. McDonnell, F.N., Baudouin, A.P., Garvey, P.M., Luxat, J.C.: CANDU Reactor Kinetics Benchmark Activity. Nucl. Sci. and Engg. 64, 95–105 (1977)
66. Modak, R.S., Gupta, A.: A Scheme for the Evaluation of Dominant Time Eigenvalues. Annals of Nuclear Energy 34(3), 213–221 (2007)
67. Mohler, R.R., Shen, C.T.: Optimal Control of Nuclear Reactors. Academic Press (1970)
68. Moiseev, N.N., Chernousko, F.L.: Asymptotic Methods in the Theory of Optimal Control. IEEE Trans. Aut. Ctrl. AC-26, 993 (1981)
69. Moore, B.C.: Principle Component Analysis in Linear System: Controllability, Observability, and Model Reduction. IEEE Trans. Aut. Contr. AC-26(1), 17–32 (1981)

70. Munoz-Cobo, L.L., Chiva, S., Escriva, A.: A reduced Order Model of BWR Dynamics with Subcooled Boiling and Modal Kinetics. Annals of Nuclear Energy 31, 1135–1162 (2004)
71. Murata, S., Ando, Y., Suzuki, M.: Design of a High Gain Regulator by the Multiple Time Scale Approach. Automatica 26, 585 (1990)
72. Naidu, D.S.: Singular Perturbation Methodology in Control Systems. Peter Peregrinus Publication, London (1988)
73. Naidu, D.S.: Singular Perturbations and Time Scales in Control Theory and Applications–an Overview. In: Dynamics of Continuous, Discrete and Impulsive Systems. Series B: Applications & Algorithms, vol. 9, pp. 233–278 (2002)
74. Naidu, D.S., Ravinder, R.: On Three–Time–Scale Systems. In: Proc. 24th IEEE CDC, FL, USA (December 1985)
75. Naskar, M., Tiwari, A.P., Menon, S.V.G., Kumar, V.: Finite Difference Approximation for Computation of Neutron Flux Distribution in a Large PHWR. In: Proc. of International Conference on Quality, Reliability and Control (ICQRC), Le Royal Meridien, Mumbai, India (2001)
76. Oberoi, S., Bandyopadhyay, B.: Robust Control of a Laboratory Scale Launch Vehicle Model using Fast Output Sampling Technique. In: Proc. 5th Asian Ctrl. Conf., vol. 1, pp. 383–391 (July 2004)
77. O'Reilly, J.: Two Time Scale Feedback Stabilization of Linear Time Varying Singularly Perturbed Systems. J. Franklin Inst. 308, 465 (1979)
78. Ott, K.O., Meneley, D.A.: Accuracy of Quasistatic Treatment of Spatial Reactor Kinetics. Nucl. Sci. and Engg. 36, 402–411 (1969)
79. Ozguner, U.: Near–Optimal Control of Composite Systems: The Multi Time–Scale Approach. IEEE Trans. Aut. Ctrl. AC-24, 652 (1979)
80. Park, Y.H., Cho, N.Z.: A Compensator Design Controlling Neutron Flux Distribution via Observer Theory. Annals of Nuclear Energy 19, 513–525 (1992)
81. Pernebo, L., Silverman, L.M.: Model Reduction via Balanced State Space Representation. IEEE Trans. Aut. Contr. AC-27(2), 382–387 (1982)
82. Phillips, R.G.: Reduced Order Modelling and Control of Two–Time–Scale Discrete Systems. Int. J. Ctrl. 31, 4 (1980)
83. Randall, D., St. John, D.S.: Xenon Spatial Oscillations. Nucleonics 16(3), 82 (1958)
84. Ravetto, P., Rostagno, M.M.: Application of the Multipoint Method to the Kinetics of Accelerator–Driven Systems. Nucl. Sci. and Engg. 148, 79–88 (2004)
85. Reddy, G.D., Bandyopadhyay, B., Tiwari, A.P.: Multirate Output Feedback Based Sliding Mode Spatial Control for a Large PHWR. IEEE Trans. Nucl. Sci. 54, 2677–2686 (2007)
86. Reddy, G.D., Park, Y.J., Bandyopadhyay, B., Tiwari, A.P.: Discrete–time Output Feedback Sliding Mode Control of a Large Pressurized Heavy Water Reactor. In: Proceedings of 17th World Congress, IFAC, Seoul, Korea, pp. 8648–8653 (2008)
87. Reddy, G.D., Bandyopadhyay, B., Tiwari, A.P., Fernando, T.: Spatial Control of a Large Pressurized Heavy Water Reactor using Sliding Mode Observer and Control. In: Proceedings of ICARCV, Hanoi, pp. 2142–2147 (December 2008)
88. Reddy, P.B., Sannuti, P.: Optimal Control of a Coupled–Core Nuclear Reactor by a Singular Perturbation Method. IEEE Trans. Aut. Contr. AC-20, 766–769 (1975)
89. Rosenbrock, H.H.: Distinctive Problems of Process Control. Chemical Engg. Progress 58(9), 43–50 (1962)
90. Safety Analysis Report - Preliminary: Physics. BARC Report No. RPDD/AHWR/80/2007 Rev 0 (2007)

91. Saksena, V.R., O'Reilly, J., Kokotovic, P.V.: Singular Perturbations and Time–Scale Methods in Control Theory: Survey 1976 – 1983. Automatica 20, 273 (1984)
92. Sannuti, P., Wason, H.: Multiple Time–Scale Decomposition in Cheap Control Problems - Singular Control. IEEE Trans. Aut. Ctrl. AC-30, 633 (1985)
93. Shamash, Y.: Stable Reduced Order Models Using Pade Type Approximation. IEEE Trans. Aut. Contr. AC-19(5), 615–616 (1974)
94. Shinkawa, M., Yamane, Y., Nishina, K., Tamagawa, H.: Derivation of Neutron Generation Time for Reflected Systems and its Physical Interpretation. Nucl. Sci. and Engg. 67, 19 (1978)
95. Shimjith, S.R., Tiwari, A.P., Bandyopadhyay, B.: A three-time-scale approach for design of linear state regulator for spatial control of Advanced Heavy Water Reactor. IEEE Trans. Nucl. Sci. 58, 1264–1276 (2011)
96. Shimjith, S.R., Tiwari, A.P., Bandyopadhyay, B.: Design of Fast Output Sampling controller for three–time–scale systems: application to spatial control of Advanced Heavy Water Reactor. IEEE Trans. Nucl. Sci. 58, 3305–3316 (2011)
97. Shimjith, S.R., Tiwari, A.P., Bandyopadhyay, B., Patil, R.K.: Spatial stabilization of Advanced Heavy Water Reactor. Annals of Nuclear Energy 38, 1545–1558 (2011)
98. Shimjith, S.R., Tiwari, A.P., Naskar, M., Bandyopadhyay, B.: Space–time kinetics modeling of Advanced Heavy Water Reactor for control studies. Annals of Nuclear Energy 37, 310–324 (2010)
99. Shimjith, S.R., Tiwari, A.P., Bandyopadhyay, B.: Three–time–scale decomposition and composite control of a nuclear reactor. In: Proc. NCVII 2009, BITS Pilani (November 2009)
100. Shimjith, S.R., Tiwari, A.P., Bandyopadhyay, B.: Coupled neutronics-thermal hydraulics model of Advanced Heavy Water Reactor for control system studies. In: Proc. IEEE INDICON 2008, IIT Kanpur (December 2008)
101. Singh, N.P., Singh, Y.P., Ahson, S.I.: An Iterative Approach to Reduced–Order Modeling of Synchronous Machines Using Singular Perturbation. Proc. IEEE 74(6), 892 (1986)
102. Sinha, N.K., Kuszta, B.: Modeling and Identification of Dynamic Systems. Van Nostrand Reinhold, New York (1983)
103. Sinha, R.K., Kakodkar, A.: Design and development of AHWR–the Indian thorium fuelled innovative nuclear reactor. Nucl. Eng. and Design 236, 683–700 (2006)
104. Sinha, R.K., Kakodkar, A.: The road map for a future Indian nuclear energy system. In: Proc. International Conference on Innovative Technologies for Nuclear Fuel Cycles and Nuclear Power, Vienna (June 2003)
105. Sinha, R.K., Kushwaha, H.S., et al.: Design and development of AHWR—-the Indian thorium fuelled innovative nuclear reactor. In: Proc. INSAC 2000: Annual Conference of Indian Nuclear Society, Mumbai (June 2000)
106. Stacey, W.M.: Space–Time Nuclear Reactor Kinetics. Academic Press, New York (1969)
107. Stammler, R.J.J., Abbate, M.J.: Methods of Steady State Reactor Physics. Academic Press, New York (1983)
108. Sharma, G.L., Bandyopadhyay, B.: Robust Controller Design for a Pressurised Heavy Water Reactor. Proceedings of IEEE Ind. Tech. 1, 301–306 (2001)
109. Sharma, G.L., Bandyopadhyay, B., Tiwari, A.P.: Spatial Control of a Large Pressurized Heavy Water Reactor by Fast Output Sampling Technique. IEEE Trans. Nucl. Sci. 50(5), 1740–1751 (2003)
110. Suda, N.: On Controllability of Neutron Flux Distribution. Jnl. Nucl. Sci. and Tech. 5(6), 271–278 (1968)

111. Sutton, T.M., Aviles, B.N.: Diffusion Theory Methods for Spatial Kinetics Calculations. Progress in Nuclear Energy 30(2), 119–182 (1996)
112. Syrmos, V.L., Abdallah, C.T., Dorato, P., Grigoriadis, K.: Static output feedback–A Survey. Automatica 33(2), 125–137 (1997)
113. Talange, D.B., Bandyopadhyay, B., Tiwari, A.P.: Spatial Control of Large PHWR by Decentralized Periodic Output Feedback and Model Reduction Techniques. IEEE Trans. Nucl. Sci. 53, 2308–2317 (2006)
114. Tiwari, A.P.: Modeling and Control of a Large Pressurized Heavy Water Reactor. PhD Thesis, IIT Bombay (1999)
115. Tiwari, A.P., Bandyopadhyay, B., Werner, H.: Spatial Control of a Large PHWR by Piecewise Constant Periodic Output Feedback. IEEE Trans. Nucl. Sci. 47(2), 389–402 (2000)
116. Tiwari, A.P., Reddy, G.D., Bandyopadhyay, B.: Design of Periodic Output Feedback and Fast Output Sampling based Controllers for Systems with Slow and Fast Modes. Asian Jnl. Ctrl., doi:10.1002/asjc.357
117. Trombetti, T.: A Nodal Approach to Stability in Nuclear Reactor Spatial Dynamics. Int. Jnl. of Energy 21(9), 1069–1079 (1983)
118. Tuttle, P.E.: Network Synthesis. John Wiley, New York (1958)
119. Wachpress, E.L., Burgess, R.D., Baron, S.: Multichannel Flux Synthesis. Nucl. Sci. and Engg. 12, 381–389 (1962)
120. Wasow, W.: On the asymptotic solution of boundary value problems for ordinary differential equations containing a parameter. J. Math. and Phys. 23, 173 (1944)
121. Werner, H.: Multimodel robust control by fast output sampling - an LMI approach. Automatica 34, 1625–1630 (1998)
122. Werner, H.: Robust control of a laboratory flight simulator by nondynamic multirate output feedback. In: Proc. IEEE CDC, pp. 1575–1580 (December 1996)
123. Werner, H., Furuta, K.: Simultaneous stabilization based on output measurement. Kybernetika 31, 395–411 (1995)

Index

CPSIA information can be obtained at www.ICGtesting.com
Printed in the USA
LVOW100254100113
315131LV00009B/142/P